# Guidebook for the Scientific Traveler

**THE SCIENTIFIC TRAVELER**

Duane S. Nickell, Series Editor

The Scientific Traveler series celebrates science and technology in America by highlighting places to visit of interest to educators, vacationers, and enthusiasts alike. Each book gives readers an introduction to the stories behind the sites, museums, and attractions related to topics like astronomy and space exploration, industry and innovation, geology and natural science. Doubling as a guidebook, each provides readers with useful and practical information for planning their own science-themed trips across America.

*Guidebook for the Scientific Traveler: Visiting Astronomy and Space Exploration Sites Across America,* by Duane S. Nickell

*Guidebook for the Scientific Traveler: Visiting Physics and Chemistry Sites Across America,* by Duane S. Nickell

# Guidebook for the Scientific Traveler

## VISITING PHYSICS AND CHEMISTRY SITES ACROSS AMERICA

**DUANE S. NICKELL**

Rutgers University Press • New Brunswick, New Jersey, and London

Sources for photos on title page and chapter openers:

Title page   Peter Ginter
Chapter 1   Dmitry Rukhlenko/Shutterstock
Chapter 2   Michael D. Brown/Shutterstock
Chapter 3   Richard Thornton/Shutterstock
Chapter 4   emin kuliyev/Shutterstock
Chapter 5   Spectral—Design/Shutterstock
Chapter 6   Ian Klein/Shutterstock
Chapter 7   iofoto/Shutterstock
Chapter 8   Corning Museum of Glass
Chapter 9   tkachuk/Shutterstock

Library of Congress Cataloging-in-Publication Data

Nickell, Duane S.
  Guidebook for the scientific traveler. Visiting physics and chemistry sites across America /
Duane S. Nickell.
        p. cm. — (The scientific traveler)
  Includes bibliographical references and index.
  ISBN 978-0-8135-4730-5 (pbk. : alk. paper)
  1. Physics—Popular works. 2. Physics—Miscellanea. 3. Chemistry—Popular works.
4. Chemistry—Miscellanea. 5. Science museums—United States. 6. Laboratories—United
States. I. Title. II. Title: Visiting physics and chemistry sites across America.
  QC24.5.N53 2010
  530.0973—dc22

                                                                    2009020405
A British Cataloging-in-Publication record for this book is available from the British Library.

Visit our Web site: http://rutgerspress.rutgers.edu

Manufactured in the United States of America

*This book is dedicated to my daughters,*
*Anna Marie Nickell and Sarah Elizabeth Nickell.*

# CONTENTS

# PREFACE

It would be difficult to overstate the influence that the sciences of physics and chemistry have had, especially throughout the last century, on both our everyday lives and the larger world. Discoveries in physics have delivered electricity to every corner of the country, ushered in the nuclear age, and placed computers and electronic gadgets at our disposal. Advances in chemistry have resulted in new synthetic materials and medicines that can save, extend, and improve our lives. This book celebrates physics and chemistry by listing, describing, and providing background information on places you can go to experience these powerful scientific disciplines. Many of these sites are conspicuously absent from traditional travel books that emphasize history, art, culture, and food but ignore science. Recent travel books listing thousands of places to visit "before you die" feature many hotels, but only a few scientific sites. This book and the other books in the Scientific Traveler series attempt to fill this void and give science its due.

I chose these sites for their scientific and historical significance. Each entry provides background information that, I hope, will make a visit much more meaningful. My main practical criterion was the site's public accessibility. Some sites are simply buildings, houses, or facilities that can be viewed only from the outside, and some sites have public tours. Because of increased security following the events of 9/11, some military and government sites require months of planning in advance of a visit. I have made personal visits to many, though certainly not all, of the sites. The visiting details provided herein were accurate at the time of writing, but the reader should certainly consult the websites for up-to-date information. I have not included directions to the sites because these can easily be found on the website or by doing a Mapquest. Although I have striven for accuracy, I'm sure that a few errors have crept into my descriptions. I apologize for these errors in advance and take full responsibility for them. I appreciate any input the reader might have that could improve any future editions of this book. Please forward any comments, corrections, and site suggestions directly to me at duane_nickell@yahoo.com.

The entries in this book fall into two types: generic and specific. In a generic entry, the background information is followed by several sites related to the topic. I used generic entries where it would be impractical and awkward to have a separate entry devoted to each site. A specific entry provides information on a single site. I do not claim that this book provides an exhaustive listing of sites related to physics and chemistry. For the ambitious scientific traveler, more sites can be identified by consulting the resources listed at the end of the book.

The reader will no doubt notice that the book contains many more sites related to physics than to chemistry. I assure you that although my training is in physics, this imbalance was not an intentional slight aimed at my chemistry colleagues. The inequality is partially explained by the fact that much chemistry is done in an industrial setting by private companies that are inaccessible to the public, whereas physics research is more often performed at large, taxpayer-funded government laboratories where public outreach constitutes part of their mission. There is another factor to consider: chemistry is often conducted on a small scale using test tubes, beakers, and glassware whereas physics is conducted on a big scale using exotic equipment like giant particle accelerators and reactors. The equipment used for physics simply makes for a more interesting site. Of course, these two factors are not unrelated; physics is done at government laboratories precisely because the equipment is too expensive for individual companies and universities to build. Chemistry, however, is relatively inexpensive. Finally, physics is a broad science encompassing all natural phenomena; at a fundamental level, everything is connected to physics. This means that a wider range of sites— sites related to energy, for example—can be justified for inclusion.

I broadly organized this book into chapters on people, places, and things. The book begins with separate chapters dedicated to physicists and chemists, the people who actually *do* the science. Then, I explore two of the main venues for scientific research—universities and national laboratories. I devote a separate chapter to the largest scientific machines ever built: particle accelerators. Chapter 6 describes sites related to the development and testing of nuclear weapons, the most historically significant manifestation of twentieth-century physics. Chapter 7 discusses energy, the physics-related issue that currently dominates newspaper headlines and political discussions. Chapter 8 details the chemical industry and related industries where chemistry plays an important role. The book closes with chapter 9, which features museums with a special emphasis on physics or chemistry.

I sincerely hope that this book will provide the reader and the traveler with a deeper understanding of how physics and chemistry is done, a deeper appreciation for the brilliant and dedicated people who do it, and a deeper awareness for how science impacts our lives and our world. Although this book is written for an adult audience, I trust that children and teenagers will accompany their parents to some of these interesting places. Maybe the visit will pique the curiosity of a young mind and open it to the possibility of a scientific career.

I have tried to include enough background scientific and historical material to keep even the armchair traveler entertained, but I hope the reader will have an opportunity to visit at least a few of the places described in this book. You need no more justification than simply the joy of being there—the intellectual pleasure one gets from standing in a place where scientific history was made. Enjoy your scientific travels!

# ACKNOWLEDGMENTS

Let me begin by thanking the Lilly Endowment for awarding me a Distinguished Teacher Creativity Fellowship which supplied me with the two ingredients I needed to successfully complete this book: time and money. The fellowship enabled me to take a semester-long sabbatical leave from my teaching duties so that I could concentrate on traveling and writing. The fellowship also provided me with funding for travel to many sites described in this book. I have discovered that, although much information can be obtained from the internet, there is simply no substitute for personally visiting a site. This is, therefore, a better book because of the support of the Lilly Endowment. In addition, I would like to thank the administration of the Franklin Township Community School Corporation for allowing me to take full advantage of the opportunities afforded by the fellowship.

Thanks to my editor, Doreen Valentine, for her continuing enthusiastic support for the Scientific Traveler series, to Robert Ehrlich for reviewing the manuscript, to Lisa Jerry for editing the manuscript, and to all the good people at the Rutgers University Press for their hard work on this book. Thanks to my aunt, Lucy Metcalf, for her hospitality while visiting some of the sites. I would like to acknowledge the late Isaac Asimov and Carl Sagan for their inspiration and influence throughout my life. Most of all, I would like to thank my parents Anna June and Carl Duane Nickell who, in addition to putting a roof over my head and food on the table, supplied me with lots of good books to read.

# Guidebook for the Scientific Traveler

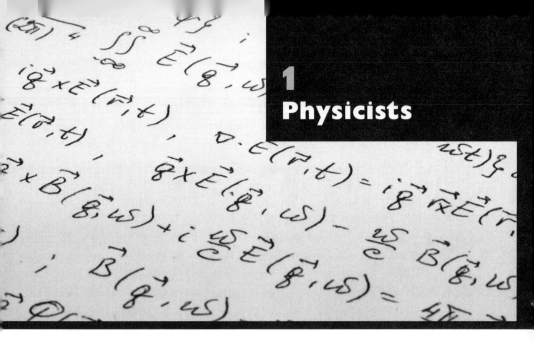

# 1
# Physicists

*The supreme task of the physicist is to arrive*
*at those universal elementary laws from which*
*the cosmos can be built up by pure deduction.*

Albert Einstein

Physicists have changed the world in profound ways. In the nineteenth century, the laws of thermodynamics formed the basis for the Industrial Revolution, and the laws governing electricity and magnetism led to lighting up a room with the flip of a switch. In the twentieth century, our understanding of the subatomic world has resulted in television, computers, and nuclear weapons. We are in the midst of an electronic revolution, but the word "electronics" is formed out the word "electron." In 1999, *Time* magazine named physicist Albert Einstein as its "Person of the Century." Some have gone even further by proclaiming the twentieth century as the century of physics.

So what, exactly, does a physicist do? Simply put, physicists try to figure out how the universe works—what makes it all go. Physicists are driven by an insatiable, childlike curiosity to understand, at its most basic level, the natural world around us. Specifically, physicists study things like motion, forces, matter, energy, light, sound, heat, electricity, and magnetism. Physics is the most fundamental science because the laws of physics govern everything. Physical laws determine how atoms interact—chemistry—and if the chemicals work together in certain, very complex ways, living organisms result—biology. Physics is the basis for chemistry, and chemistry is the basis for biology.

The first true physicist was Galileo Galilei, who, for the first time, used data obtained from experimentation to develop mathematical laws that describe natural phenomena. He established mathematics as the concise, unambiguous language of physics. As physics matured in the late nineteenth and early twentieth centuries, a division of labor among physicists evolved. Physicists who liked to get their hands dirty by tinkering with laboratory equipment and who excelled at devising experiments to test ideas became known as experimental physicists. Perhaps the greatest experimental physicist ever was the English physicist Michael Faraday, who discovered many basic laws of electricity and magnetism. Physicists who like to play with equations, those who fill reams of paper with their incomprehensible scribbling, are called theoretical physicists. Theoreticians use mathematics to explore possible ways the universe might work. Albert Einstein is the quintessential theoretical physicist who, by using his mind as a laboratory, gave us new conceptions of space and of time. In essence then, theoreticians are the dreamers who generate ideas about how the universe might work, and experimentalists are the realists who devise tests to see if these ideas correspond to how the universe actually does work.

The explosive growth in scientific knowledge throughout the twentieth century means that no single physicist can possibly know all there is to know about the discipline. Thus, physicists have to focus on one particular area of physics. Just as medical doctors specialize in certain diseases or parts of the body, physicists likewise specialize in areas such as solid-state physics, particle physics, nuclear physics, or astrophysics. So, if you go up to a friendly looking physicist and ask him or her what kind of physicist he or she is, then you might get this kind of answer: "I'm an experimental nuclear physicist," or "I'm a theoretical particle physicist."

Roughly forty thousand professional physicists work in the United States. About half are employed by the private sector, 25 percent by academia, and 20 percent by the government. The entries below give biographical sketches of a few famous American physicists along with a place or, in some cases, multiple places you can visit to make a connection to these great scientific minds. Most sites are simply houses where they once lived, but a couple are actual museums. Three of the houses (the Millikan, Compton, and Fermi houses) sit along the same street in the Hyde Park neighborhood of Chicago. The entries are arranged in chronological order according to the physicist's date of birth.

## Benjamin Franklin

Benjamin Franklin is rightly famous for being a Founding Father of our country, but he was also a scientist of the first rank. His legacy is beautifully captured by the following epigram composed by the French economist Jacques Turgot: "He snatched lightning from the sky and the scepter from tyrants." A product of the Enlightenment, Franklin believed the human mind could make sense of the natural world and was passionately curious about scientific phenomena. His inquisitiveness compelled him to investigate everything from ocean currents to thermodynamics, from agriculture to the aurora. But Ben is best known for his electrical explorations.

In the eighteenth century, electricity was on the scientific frontier, and Franklin had to create new language to discuss the results of his experiments. Terminology that Franklin himself introduced into the study of electricity includes positive, negative, plus, minus, charge, condenser, conductor, battery, and armature. His most important theoretical insights included an explanation of what happened when materials like glass and wool were rubbed together. According to Franklin, electricity is not actually created in this process; rather, the glass simply extracts the "Electrical Fire" from the silk and absorbs it. The amount of electrical fire lost by the silk equals the amount gained by the glass. Franklin incorrectly thought of electricity as a fluid but correctly claimed that the fluid could be neither created nor destroyed; electricity was simply transferred from one place to another. This idea later evolved into the principle of the conservation of electric charge, one of the most basic laws governing electricity and magnetism.

Of course, Franklin is best remembered for his experiment with lightning. In Franklin's time and before, lightning was an awesome phenomenon shrouded in mystery. Ancient peoples believed that lightning bolts were spears that the gods hurled at their enemies. In Franklin's time, lightning was seen as God's way of punishing sinners, and grown men cowered under their beds in superstitious fearfulness during thunderstorms. Franklin was not the first to suggest that lightning might be a giant electrical discharge; indeed, many electrical experimenters had noted similarities between lightning and sparks of electricity. Franklin's breakthrough, however, was twofold: first, he described an experiment that could decide whether thunderclouds were electrified; and second, if thunderstorms were electrical, then he suggested that tall grounded rods could protect structures from lightning strikes.

Franklin's electrical experiments and explanations were described in a series of letters composed from 1747 through 1750 and sent to Peter Collins, a fellow of London's Royal Society. In 1751 the Royal Society published the letters in the form of a pamphlet, "Experiments and Observations on Electricity, Made at Philadelphia in America." The pamphlet proposed the following experiment: a sharp iron rod, twenty to thirty feet tall could be positioned vertically on a wooden platform to insulate it from the ground; if, during a passing storm, sparks could be drawn from the rod, then that would prove that the thunderclouds are electrified.

On May 10, 1752, a French naturalist actually did the experiment, drew sparks from the rod, and proved the electrical nature of thunderstorms. When the results were reported in Paris, with proper acknowledgment to Franklin, it created a sensation. About a month later, before he knew about the French experiment, it occurred to Franklin that he could get closer to the thunderclouds simply by flying a kite. There is some historical debate as to whether Franklin actually performed the kite experiment. Franklin himself never wrote about it. A written account wasn't published until fifteen years later in chemist Joseph Priestly's *History and Present Status of Electricity*, although presumably Franklin provided Priestly with the details. If Franklin did conduct the experiment, then he certainly didn't do it in the way it is usually portrayed. That is, he didn't fly the kite in a thunderstorm and wait for lightning to strike it—that would have killed Franklin as it did several others who tried to repeat the experiment. Franklin's writings indicate that he was well aware of the dangers that electricity posed. Rather than waiting for lightning to hit the kite, he most likely flew the kite early in the storm before lightning approached. Franklin saw loose strands of thread stand out from the kite string, a clear sign that it was electrified. He then placed his knuckle close to the key that was attached to the string and drew an electrical spark. It is also possible that he attached the string directly to a Leiden Jar (a device used to store electrical charge, the equivalent of what we would today call a capacitor), electrified it, and then drew sparks from the jar.

Armed with the knowledge that lightning was electrical in nature, Franklin now turned his attention to preventing lightning damage. During his experiments, he had noticed that electricity seemed to flow more readily from sharp points as opposed to a rounded surface. After installing a

The Benjamin Franklin National Monument at the Franklin Institute in Philadelphia.

prototype lightning rod on his own house, Franklin published a method for protecting structures from lightning in the 1753 issue of *Poor Richard's Almanac*. The main purpose of a lightning rod, contrary to popular belief, is not to attract lightning to keep it from hitting the structure. Instead, lightning rods reduce the chance that lightning will strike by preventing the build-up of electrical charge on the structure. If the lightning does strike, it will be attracted to the rod and dumped into the ground.

Franklin's electrical discoveries made him world famous. He had demonstrated mankind's ability to understand nature through science and control nature through technology and invention. Franklin's scientific reputation, which at the time rivaled that of Sir Isaac Newton, served him well in his later roles as a colonial envoy to England and ambassador to France. Franklin was more interested in the practical applications of science than in its theoretical underpinnings and became a prolific inventor. He modified the stove so that it would use less wood and deliver more heat. He designed a three-wheeled clock that was much simpler than most clocks of the time. As postmaster, he crafted an odometer that could be attached to the wheels of carriages to measure distances. The resulting data was used to draw up more efficient postal routes. He created an extension arm to help him reach books on the upper shelves of his library; this device consisted of a long pole with two "fingers" on the end that could be opened and closed by pulling on a cord. An excellent swimmer, Franklin devised swim fins in the shape of a lily pad that could be worn on the hands to increase speed. To economize the use of candles, Franklin suggested the idea of Daylight Savings Time. Perhaps his most famous invention was bifocal glasses. A contemporary joked that Franklin, who was something of a lady's man, invented bifocals so he could look at women across the room while still keeping an eye on the ones nearby. Franklin's explanation was more innocent: he simply didn't like carrying two pairs of glasses when traveling—one for reading and the other for taking in the view of the countryside. Franklin, declining to seek patents on any of his inventions, explained: "As we benefit from the inventions of others, we should be glad to share our own . . . freely and gladly."

## Visiting Information

There are a number of sites, most in Philadelphia, where the scientific traveler can make a connection with Benjamin Franklin. The site of Franklin's birth is in Boston at 1 Milk Street, opposite the Old South Meeting Place where he was baptized. The site is on Boston's popular walk, the "Freedom Trail." Unfortunately, the house was destroyed by fire in 1811. The front of the building that now stands there has the words "Birthplace of Franklin" carved above a second-story window along with a bust of Franklin. A statue of Franklin can be found a block away on School Street in front of the Old City Hall. Franklin's father, Josiah, ran a soap and candle shop at Hanover and Union streets.

In Philadelphia, Franklin Court, located on Market Street between Third and Fourth streets, is the main Franklin-related site. Walk through the arch into the court to see a steel-frame "ghost structure" that outlines the spot where Franklin's house, the only one he ever owned, once stood. The original house was torn down in 1812 to make way for commercial development, and no historical records show what the exterior of the house looked like. Historians do know that it was a square, brick house, thirty-four feet on each side, three stories tall, with three rooms on each floor. A kitchen occupied the cellar, and chimneys were located on both sides of the house. Franklin remarked with satisfaction that it was "a good house contrived to my mind." The house was built between 1763 and 1765, but, owing to his ventures abroad, he lived here only a few years of his life. When he finally returned to the house at age eighty, he expanded it by half to include a library, two more bedrooms, two garrets, and a place to store wood. Today, you can look down into the site to see the foundation, wells, and privy pits.

Below the court is a museum with artifacts, paintings, and reproductions of his inventions including a glass harmonica, a Franklin stove, and a swim fin. A bank of phones lets you listen to brief testimonials about Franklin from Washington, Mozart, and many others. You can also watch a twenty-two-minute film, *The Real Ben Franklin*.

Along Market Street are restorations of five buildings, three of which Franklin built as rental properties. These buildings include U.S. Postal Service Museum, a post office, an eighteenth-century printing office and bindery, the office of the *General Advertiser*, a newspaper published by Franklin's grandson, and an architectural exhibit showing how Franklin tried to make buildings more fire-resistant.

At Fifth Street and Vine is a sculpture by Isamu Noguchi titled *Bolt of Lightning . . . A Memorial to Benjamin Franklin*. The 60-ton, 101-foot-tall stainless steel sculpture depicts a kite at the top, a bolt of lightning, and a key at the base. Noguchi first proposed the sculpture in 1933, but the design was considered too radical. Fifty years later, artistic sensibilities had changed, and Noguchi himself picked the location. A huge bust of Franklin can be seen outside the Engine 8 firehouse at Fourth and Arch streets. Why at a fire station? It turns out that Franklin founded the first professional fire company in the New World in Philadelphia.

The Benjamin Franklin National Memorial is located in the rotunda of the Franklin Institute Science Museum and features a 20-foot-tall seated statue of Franklin sculpted in marble by James Earle Fraser between 1906 and 1911. The statue, positioned on a pedestal of white marble, is the focal piece of the Memorial Hall, modeled after the Pantheon in Rome. Memorial Hall houses some of Franklin's personal possessions, such as his composing table and several original publications. The Franklin Institute is not to be missed, and I describe it in detail in chapter 9 of this book. The institute's Franklin Gallery holds a variety of artifacts including one of Franklin's lightning rods, a glass harmonica, and an electrostatic machine that he used for some of his experiments.

The only remaining intact Franklin home is located in London, England, near Trafalgar Square at 36 Craven Street. The "Benjamin Franklin House" is now a museum and educational facility. Franklin lived there between 1757 and 1775 while he was mediating unrest between Britain and the colonies.

Franklin is buried in the Christ Church Burial Ground along with four other signers of the Declaration of Independence. The grave is visible through an iron gate at the southeast corner of Fifth and Arch streets, around the corner from Franklin Square. His tombstone reads simply, "Benjamin and Deborah Franklin, 1790." Tossing a penny on the grave supposedly brings good luck.

> Website:  www.nps.gov/inde/
> Telephone:  215–965–2305
> (Independence National Historical Park)

## Count Rumford (Benjamin Thompson)

Benjamin Thompson was born into a moderately wealthy farm family in Woburn, Massachusetts, in 1753. Although his father died when he was very young, Benjamin was well cared for and received his only formal education at the village school. At age thirteen, he was apprenticed to a merchant in Salem, socialized with well-educated people, and developed an interest in science. At sixteen, he conducted experiments on the nature of heat and discussed his results with his friend and neighbor, Loammi Baldwin (considered by many to be America's first engineer). Thompson and Baldwin occasionally attended lectures on science given by Professor John Winthrop at Harvard.

At age nineteen, Thompson became a schoolmaster in Rumford (now Concord), New Hampshire. There, he met, courted, and married a wealthy, older widow named Sarah Rolfe. When the American Revolution broke out, Thompson, now a rich property owner as a result of his marriage, sided with the British and became a spy. His activities raised suspicion among the rebels, and he was twice charged with "being unfriendly to the cause of liberty." Although he was acquitted in both cases for lack of evidence, he was feeling the heat so, when the British evacuated Boston in March 1776, Thompson retreated with the redcoats, deserting his wife and young daughter, whom he never saw again.

Thompson worked alongside the British during the Revolutionary War and performed experiments concerning the explosive force of gunpowder. The research establishing Thompson as a skilled scientist, was eventually published in the *Philosophical Transactions of the Royal Society*. When he moved to England at the end of the war, he was elected as a member of the prestigious organization. After a few years hobnobbing with London's high society, Thompson was offered a high-ranking position in the Bavarian court in Munich. He obtained permission from the British government to accept the position and, before leaving England, was knighted by King George III.

Sir Benjamin Thompson spent the next eleven years in Munich where one of his main responsibilities was reinvigorating the Bavarian army, whose soldiers were suffering from low morale. Thompson gave the soldiers a pay raise and guaranteed a free education for themselves and their children. When not in training, he used the army to carry out ambitious public works projects, the most famous of which is Munich's English Garden. He also wanted the soldiers to have better and warmer uniforms. To this end, he undertook a study of the thermal properties of various kinds of cloth. This resulted in the discovery of heat transfer by convection and the invention of thermal underwear. When Thompson finished designing the uniforms, he needed someone to actually make them, but couldn't convince any Bavarian clothing manufacturers to take the job. To solve the problem, Thompson resorted to an ingenious bit of social engineering.

At the time, Munich was plagued with swarms of beggars. On New Year's Day, 1790, Thompson ordered the army to round up and arrest every beggar in the city. But instead of putting them behind bars, he sent them to a workhouse where he trained the beggars to piece together the uniforms. In exchange, the beggars got food, shelter, and an education. To ensure the efficient operation of the facility, Thompson came up with a recipe for a nutritious soup, invented the wax candle to replace the overly smoky tallow or fat candles, and designed a new type of stove that would reflect heat into the room while carrying away the smoke. In recognition of his efforts, Thompson was made a count of the Holy Roman Empire. For his title he chose the name of the little New Hampshire town in which he once lived.

Count Rumford is best known for the scientific contribution that resulted from his duty to oversee the making of cannons for the Munich arsenal. He was amazed at the amount of heat generated in the boring of cannon. To study this phenomenon, he used a blunt borer to maximize the heating effect and found that he could boil large quantities of water by using the heat produced. Indeed, an almost inexhaustible supply of heat seemed to be available. This fact appeared to contradict the prevailing theory that viewed heat as caloric, a fluid that flowed into and out of materials. According to the caloric theory, the heat produced by the boring of the cannon developed because the boring tool was squeezing caloric out of the materials that the cannon and bore were made of. But Rumford argued that, just as a jug could hold a finite amount of water, the cannon and bore held only a finite amount of caloric; yet, as long as the boring process continued, heat was always available. Rumford concluded that the heat was not in the metal; heat was being generated by the motion of the bore. In 1798, Count Rumford presented a paper to the Royal Society entitled "Enquiry Concerning the Source of Heat which is Excited by Friction" in which he put forth the idea that heat was a form of motion. It took a while for this idea to take hold in the scientific community, but it was an important stepping stone toward one of the great ideas in all of science, the conservation of energy.

Rumford's story now takes a remarkably ironic twist. The use of the word "caloric" had been introduced by the great French chemist Antoine Lavoisier who had been beheaded during the French Revolution. After Count Rumford returned to England, he moved to Paris where he had an affair with Lavoisier's wealthy widow, which ended in a troubled two-year marriage. (Rumford's American wife had died in 1792.) In 1807, Rumford retired outside Paris to the village of Auteuil where he died suddenly in 1814 at age sixty-two. He is buried in a small cemetery there.

## Visiting Information

Count Rumford was born in a house in Woburn, Massachusetts, on March 26, 1753. The house had been built by his grandfather, Captain Ebenezer Thompson, and his parents had lived in it since they were married. The house is furnished with Rumford's baby crib and other artifacts. It also serves as a museum with several laboratory models that demonstrate Rumford's ideas about heat and a diorama of his cannon boring experiment. A library holds essays and biographies about Rumford and a copy of a portrait of Rumford by Gainsborough can be viewed. The house is located at 90 Elm Street in North Woburn, about ten miles northwest of Boston. Hours are from 1:30 P.M. to 4:30 P.M. on

Website:  www.woburnhistoricalsociety.com
Telephone:  781–287–0260

weekends and admission is free. The caretakers of the house give narrated tours. Elsewhere in town, a statue of Rumford stands in front of the Woburn Public Library located at 45 Pleasant Street. The statue is a copy of the original that stands in Munich.

# Joseph Henry

Joseph Henry was born into a poor, working-class family of Scottish immigrants in Albany, New York, in 1797. His father was an alcoholic day-laborer who died when Joseph was only eight. The family's dire financial situation forced Joseph's mother to send him to live with his grandmother in Galway, about forty miles from Albany. There, he attended school, worked in a general store, and, at age thirteen, became an apprentice to a watchmaker. The next year he moved back to Albany, where he worked during the day and took night classes at the Albany Academy, for which he had been granted free tuition. Joseph's first love was the theater, and he almost became a professional actor. But at age sixteen, his interest in science was sparked when he read *Popular Lectures on Experimental Philosophy, Astronomy, and Chemistry.* Later in his life, Henry gave to his only son a copy of the book with this inscription: "This . . . opened to me a new world of thought and enjoyment; fixed my attention upon the study of nature, and caused me to resolve at the time of reading it that I would immediately devote myself to the acquisition of knowledge." Henry supported himself as a teacher and private tutor, then a surveyor and engineer. In 1826, he accepted a position as a professor of mathematics and natural philosophy (physics) at the Albany Academy. Before starting his new job, Henry learned about a new device, called an electromagnet, consisting of bare wire coiled loosely around a piece of iron. When the wire was connected to a battery, the iron became a magnet. Henry improved the design of electromagnets by using insulated wire wrapped in tight layers around the iron. (According to legend, Henry used silk from his wife's petticoats to insulate the wire.) To the amazement of his students, Henry's electromagnets could lift objects weighing more than a ton.

The strong magnetic fields produced by Henry's electromagnets led to his most important scientific discovery, a property of electrical circuits called self-inductance. The self-inductance of a circuit causes it to resist a change in the circuit's current. Today, the "Henry" is the unit for inductance. Henry was also codiscoverer of mutual induction (more generally referred to as "electromagnetic induction"), the principle behind electrical generators and transformers. Indeed, our modern electrical civilization is based on electromagnetic induction. Henry had an unfortunate bad habit of being slow to publish his results. Meanwhile in England, Michael Faraday was doing similar work, published a paper before Henry, and is therefore usually given more credit for the discovery.

Joseph Henry (1797–1878).

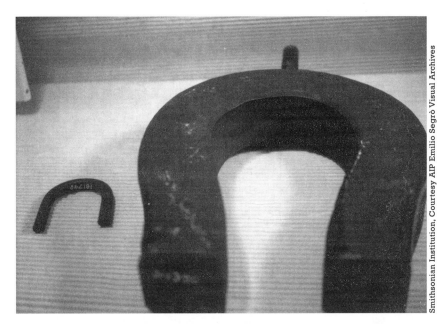

An electromagnet made by Joseph Henry in 1831.

In 1831, Henry wrote a paper describing one of the first machines to use electromagnetic forces to cause motion. The device consisted of an electromagnet that could rock up and down like a see-saw. The ends of the electromagnet curved downward to face the poles of two permanent magnets mounted vertically. Two pairs of wires extended in opposite directions from the electromagnet to the terminals of two batteries. The rocking motion was produced when the wires made contact with the battery and reversed the polarity of the electromagnet with every swing. Henry's "philosophical toy," as he called it, was one of the earliest ancestors of the modern direct current motor.

Henry's electromagnetic explorations earned him an international reputation and an invitation to join the faculty of what was then the College of New Jersey, now Princeton University. The position gave Henry a little more money and, free of a heavy teaching load, a lot more time to pursue his research. While at Princeton, Henry strung a wire from his lab to his house and devised a signaling system based on electromagnets that enabled him to send messages (sometimes to order lunch) to his wife. The principle was the same as that of the magnetic relay that Samuel Morse later applied to the invention of the telegraph.

For more than fifteen years Henry remained at Princeton, where he taught classes in physics, chemistry, geology, astronomy, and architecture. He continued his research on electromagnetism but branched out into the areas of astronomy, sound, capillary action, phosphorescence, and ballistics. He was, by all accounts, an outstanding teacher, who often said that he wanted to teach his students not merely facts but rather a way of thinking and learning that could be applied to the wider world.

Henry's career took a different direction in 1846 when an Englishman named James Smithson bequeathed his fortune for the purpose of founding an institution dedicated to "the increase and diffusion of knowledge among men." The U.S. Congress established the Smithsonian Institution and formed a search committee to seek the best candidate to serve as the institution's first secretary. The job was offered to Henry, and he accepted the position, although he recognized that his days as a scientific researcher had come to an end. In addition to organizing the new institution, Henry became the chief scientific advisor to the president, the cabinet, and the Congress; moreover, he forcefully advocated for the advancement of science. Regarding the purpose of the institution, Henry said: "The most prominent idea in my mind is that of stimulating the talent of our country to original research . . . in which it has been most lamentably difficient [sic] . . . to pour fresh material on the apex of the pyramid of science, and thus to enlarge its base."

Shortly after becoming secretary, Henry laid the foundation for what would eventually evolve into the National Weather Service. He enlisted the help of hundreds of volunteer weather watchers scattered across the entire North American continent. Each volunteer received detailed instructions, instruments, and standardized forms on which to record the data. The volunteers submitted monthly reports that included observations of temperature, pressure, wind, humidity, cloud conditions, and precipitation. Henry also recognized that the telegraph could be used to transmit local weather data to the Smithsonian to warn of approaching storms. Under Henry's direction, the Smithsonian Institution became a driving force behind the direction and improvement of American science. Joseph Henry died in 1878 at age eighty.

## Visiting Information

From 1837 until 1848, Joseph Henry designed and lived in a house located on the campus of Princeton University, and it was to this house that Henry sent telegraphic messages to his wife. The house is located on the north end of campus along Nassau Street just north of Chancellor Green. The house has been moved three times to make way for other campus buildings. The original location was on the south side of what is now Stanhope Hall. The house, the official residence of various university deans, is not open to the public. Elsewhere on campus, some pieces of Henry's laboratory equipment, including several of his electromagnets and an early motor, are on display in the lobby of Jadwin Hall. See the entry on Princeton University in chapter 3 for more details.

Other sites related to Joseph Henry include Albany's first Presbyterian Church, where Henry was baptized. A stained-glass window shows Henry demonstrating an electrical device to a class of students. The captions at the bottom of the left and right panels read "Master Scientist" and "Devout Christian." In Washington, D.C., a statue of Henry stands in front of the Smithsonian's original "Castle" building on the National Mall. A bronze portrait of Henry is one of sixteen in the Jefferson Building of the Library of Congress. Henry is buried in Oak Hill Cemetery on the northwest side of the city.

## Robert Millikan

Born in 1868 into the large family of an Iowa preacher, Robert Andrews Millikan, who won the Nobel Prize in physics, became one of America's preeminent physicists of the twentieth century. Millikan's interest in physics began when, at the end of his sophomore year at Oberlin College, his Greek professor asked him to teach introductory physics the next fall. Nobody else at Oberlin knew enough about the subject to teach it. When Millikan explained that he didn't know anything about physics either, the professor replied that anyone who could do as well as Millikan had done in his Greek class could teach physics. With the stipulation that the professor would have to take responsibility for the outcome, Millikan agreed to give it the old college try. Millikan spent most of his summer vacation teaching himself the subject. In the fall, he launched himself with gusto into his teaching assignment. Millikan later said: "I doubt if I have ever taught better in my life than in my first course in physics in 1889. I was so intensely interested in keeping my knowledge ahead of that of the class that they may have caught some of my own interest and enthusiasm."

The athletic Millikan considered a career in physical education, but a professor convinced him to continue his education at Columbia University where he earned a doctorate in physics in 1895. Following the custom for young American scientists of his generation, he spent the next year studying in Europe at the universities of Berlin and Göttingen. In 1896, the well-known physicist Albert Michelson offered Millikan a teaching assistantship at the newly established Ryerson Laboratory at the University of Chicago. Millikan accepted and, during his early career, made his mark as an outstanding classroom teacher and prolific textbook writer. The elementary physics textbooks that Millikan wrote, in collaboration with others, educated a generation of physics students. Like other large research institutions, the University of Chicago valued scientific research more than teaching excellence. At the time, most academic physicists became full professors at a median age of thirty-two, but Millikan wasn't promoted to the rank of associate professor until age thirty-eight. As a result, Millikan decided to stop writing textbooks and devote himself fully to his research.

His first project was to accurately determine the electric charge on an electron. Previous attempts had measured the charge by using clouds of water droplets. Millikan's major improvement was to use oil instead of water. The water droplets evaporated quickly whereas the oil drops could be observed for a much longer time. In Millikan's now famous "oil-drop" experiment, a fine mist of oil was sprayed into a chamber above two parallel plates attached to a source of electricity. As the drops shot out of the nozzle of the atomizer, they gained an electric charge. Some drops fell through a hole in the top plate into a region between the plates. The region was lit from the side by a high-intensity light, and the drops twinkled like stars when viewed through a microscope. By observing the falling drops, the mass of the drops could be calculated. Then, an electric field was established between the plates and adjusted until the weight of the drop was balanced by an upward electrical force. By equating the two forces, an experimenter could calculate the electrical charge on the drop. Millikan showed that the charge on the droplets was always a whole number multiple of $1.592 \times 10^{-19}$ Coulombs, the charge on a single electron. Millikan's result is only slightly off from today's accepted value of $1.602 \times 10^{-19}$ Coulombs. But his measurement had broader significance than merely refining a number. It also showed that electrons were fundamental particles and provided convincing evidence supporting the Bohr model of the atom. After publishing his results in 1910, Millikan was promptly awarded a full professorship.

Millikan's next research project was a decade-long set of experiments designed to test the validity of Einstein's equation for the photoelectric effect. (The photoelectric effect occurs when light shines on a metal surface and ejects electrons. Einstein explained this phenomenon by saying that the light consisted of little particles of energy, later known as photons.) Millikan was convinced that Einstein's equation was wrong because a mountain of evidence seemed to confirm the wave model of light. Millikan was surprised when his experiments confirmed every aspect of Einstein's photoelectric equation. Nevertheless, his data yielded the first direct measurement of Planck's constant, a fundamental constant in nature and the linchpin of quantum mechanics. For his measurement of the charge on the electron and his work on the photoelectric effect, Millikan was awarded the 1923 Nobel Prize in physics.

In 1921, Millikan left the University of Chicago to become the chairman of the executive council of the newly established California Institute of Technology. In effect, Millikan was president of Caltech, although he declined the title. During his Caltech years his scientific work focused on radiation coming from outer space, which he dubbed "cosmic rays."

Later in his life, Millikan's religious upbringing led him to lecture and write about reconciling science and Christian faith. More controversial is the fact that Millikan, a Social Darwinist, served on the board of the Human Betterment Foundation, a group that promoted eugenics research. During the 1940s, the foundation was actively documenting the medical sterilization of mentally ill people. Millikan once boasted that his hometown of San Marino was "the westernmost outpost of Nordic civilization . . . [with] a population which is twice as Anglo-Saxon as that existing in New York, Chicago, or any of the great cities of this country." Millikan died of a heart attack in 1953 at the age of eighty-five and is buried in the "Court of Honor" at Forest Lawn Memorial Park Cemetery in Glendale, California.

## Visiting Information

From 1907 to 1921, Millikan lived in a large, three-story brick Prairie School style house in Chicago along with his wife, Greta, and their three sons, Clark Blanchard (who became a prominent aerodynamic engineer), Glenn Allen, and Max Franklin. The house is located at 5605 Woodlawn Avenue in the Hyde Park community of Chicago. This private residence has been designated as a National Historic Landmark.

# Albert Einstein

Albert Einstein lived in Princeton, New Jersey, from 1935 until his death in 1955. When the Einsteins first arrived in Princeton in 1933, they rented a house. Then, in the summer of 1935, the house down the block at 112 Mercer Street went up for sale. The home is a modest two-story white clapboard structure with a big front porch and a little front yard. The house is a reflection of the man—simple and unpretentious with a slightly disheveled look to it. Einstein lived here with his wife Elsa, his stepdaughter Margot, and his secretary Helen Dukas. Through the years, they kept company with a menagerie of pets, including a parrot named Bibo, a cat named Tiger, and a white terrier named Chico who occasionally had unfriendly encounters with the mailman. Inside, the living room was dominated by massive German furniture. Helen Dukas converted a small library on the ground floor and used it as an office where she could sort through all the letters and answer the all the telephone calls. Einstein had a study on the second floor with a picture window that overlooked the backyard garden. Built-in bookcases climbed from floor to ceiling. The walls were graced with pictures of Einstein's scientific heroes: Isaac Newton, Michael Faraday, and James Clerk Maxwell, along with a political hero, Mahatma Gandhi. The only award on display was a framed certificate of membership in the Bern Scientific Society. A large wooden table in the center of the room served as a dumping ground for his papers and pencils. Einstein usually worked here in the afternoons sitting in an easy chair scribbling an endless stream of equations onto a pad of paper balanced on his lap.

So how did Einstein wind up in Princeton? Einstein was a professor in Berlin when Hitler came to power in 1933; Einstein, however, was Jewish and the Nazis, wasting no time in dismissing relativity as an erroneous Jewish theory, threw Einstein's books on their bonfires. Einstein decided to leave Germany forever and immigrate to the United States. He was drawn to Princeton to work at the newly established Institute for Advanced Study, a "think-tank" where he would be paid a generous salary but would be unencumbered by any teaching or research duties. He was completely free to follow his own intellectual desires—the perfect place for the free-spirited Einstein. During his time at the institute, Einstein worked on a unified field theory, a theory that would combine the forces of gravity and electromagnetism under one mathematical framework. He never achieved this elusive goal. However, some of the world's leading physicists work at

AIP Emilio Segrè Visual Archives, Brittle Book Collection

Albert Einstein with Professor Schilpp in his office at Princeton.

the institute today, and they are continuing Einstein's quest for the ultimate theory.

Einstein died in 1955 at the age of seventy-six. His disliked the cult of personality that had formed around him and forbade any type of funeral service. His body was cremated in Trenton on the afternoon of his death, and his ashes were scattered in the Delaware River before the rest of the world knew he was gone. He specifically requested that his house not be turned into a museum. After his death, his secretary Helen Dukas and his stepdaughter Margot continued to live in the house for more than thirty years. (His wife Elsa had died a short time after they moved into the house.)

Photograph by Alan Richards, AIP Emilio Segrè Visual Archives

Einstein's Institute for Advanced Study office, as it appeared at the time of his death in 1955.

When Margot died in 1986, the house became the property of the institute. Institute officials used the house to help entice a distinguished professor to come to Princeton. That must have been some sales pitch: "You come work for us, and you can live in Einstein's house." It worked—the professor lives here still.

### Visiting Information

The house at 112 Mercer Street is a private residence. It is not open to the public. However, the Bainbridge House at 158 Nassau Street, home of the Princeton Historical Society, has a collection of furniture that had furnished the house on Mercer Street. A few items from the collection are on display. As the city's unofficial visitor's center, the Bainbridge House distributes free self-guided walking tour brochures and pamphlets. Guided historical walking tours are offered on Sunday afternoons at 2:00 P.M. for $7. The Bainbridge House is open on Tuesday through Sunday from noon until 4:00 P.M. The Institute for Advanced Study is about a mile from the center of town on Einstein Drive. The buildings are not open to the public, but you can at least park your car and take a look. The Institute Woods are open to the public so feel free to take a leisurely stroll. Several Einstein-related sites are included on the walking tour of Princeton described in chapter 3. Einstein's summer home is located in Cutchogue (formerly known as Peconic), on Long Island. From here he wrote the famous letter to President Franklin D. Roosevelt regarding the atomic bomb. (The original letter resides in Roosevelt's archives at his home in Hyde Park.) The cottage is located at 1255 West Cove Road. Einstein kept his sailboat in Horseshoe Cove and sailed it around Cutchogue Harbor in Little Peconic Bay.

Another site all admirers of Albert Einstein will want to visit is the Einstein Memorial in Washington, D.C. Whenever I'm in Washington, I, like a pilgrim to Mecca, am inexorably drawn to this delightful memorial. Of all the monuments and memorials adorning our nation's capital, this is my personal favorite. Tucked away in a tranquil grove of elm and holly trees on the grounds of the National Academy of Sciences, the memorial is a twelve-foot-tall bronze statue of Albert Einstein, sitting on the second of three semicircular steps of white granite dug from the hills of Mount Airy, North Carolina. This bronze Einstein is dressed simply, as he often was in real life, in a sweatshirt and sandals. Sculpted by Robert Berks, the piece reminds me of the paintings of Vincent Van Gogh, where the artistic style allows a coherent whole to emerge from a myriad of individual brushstrokes. (Berks also sculpted the bust of John F. Kennedy that graces the Kennedy Center for the Performing Arts.) In sharp contrast to the formality of most of Washington's monuments, this sculpture exudes a feeling of relaxed informality. It captures the essence of Einstein's personality: unassuming and unimpressed with his own importance.

In his left hand, Einstein holds a tablet with equations that summarize three of his most important scientific achievements. The top equation is that for general relativity, the theory that introduced a new conception of gravity; the middle equation describes the photoelectric effect, a phenomenon that Einstein explained by envisioning light as behaving like little bundles of energy called "quanta" (for this explanation Einstein won the 1921 Nobel Prize in physics); the third equation is the famous $E = mc^2$ formula that sets forth the equivalence of matter and energy. Einstein seems to invite the visitor to climb into his lap and ponder the equations. Go ahead . . . there are no security guards here to chase you away. I'm sure Einstein would have liked that!

At Einstein's feet is a twenty-eight-foot-diameter circular map of the sky made of emerald-pearl granite from Norway. The 2,700 metal studs

Jason Maehl/Shutterstock

The Albert Einstein Memorial in Washington, D.C.

embedded in the granite represent the positions of the planets, stars, and various other celestial objects at noon on April 22, 1979, the day the memorial was dedicated in celebration of the centennial of Einstein's birth. The size of the studs corresponds to the apparent brightness of the objects as seen from the earth. Along the back of the steps are the following three quotations that capture Einstein's love of freedom, nature, and truth:

> As long as I have any choice in the matter, I shall live only in a country where civil liberty, tolerance, and equality of all citizens before the law prevail.

> Joy and amazement of the beauty and grandeur of this world of which man can just form a faint notion . . .

> The right to search for truth implies also a duty; one must not conceal any part of what one has recognized to be true.

Website: www.princetonhistory.org
www.ias.edu
(Institute for Advanced Study)
www.nasonline.org.
Click on "About the NAS"
and then "NAS Building."
Telephone: 609–921–6748 (Bainbridge House)

The Einstein Memorial is on the grounds of the National Academy of Sciences just across the street from the Vietnam Memorial. While you are there, you might want to take a look inside the National Academies building, which features some neat art and architecture. The website has details.

## Robert Goddard

In January 1920, the Smithsonian Institution published a paper, "A Method of Reaching Extreme Altitudes," in which the author, physicist Robert Goddard, argued that rockets were a viable way of lifting weather-recording instruments to higher altitudes than sounding balloons. After laying out, for the first time, the mathematical details supporting the idea, Goddard wrapped up the paper by speculating that a big enough rocket with a powerful enough fuel might even make it to the moon. The press picked up on this last part and ridiculed the idea. The *New York Times* condescendingly explained that a rocket could not possibly work in outer space because there was no atmosphere out there to push against. Goddard, they said, obviously lacked "the knowledge ladled out daily in high schools." Others derisively called Goddard "moon-rocket man" and "moony."

Of course, the editors of the *Times*, not Goddard, lacked a basic understanding of Newton's Third Law of Motion: "For every action there is an opposite and equal reaction." Goddard proved his critics wrong by performing a demonstration in which a rifle was suspended by wires in a glass enclosed vacuum chamber. When the rifle was fired, the bullet went one way, and the rifle kicked back in the opposite direction, which proved that the bullet didn't need air or anything else to push against in order to propel the rifle in the opposite direction. The *New York Times* finally got around to issuing a retraction when the *Apollo 11* astronauts landed on the moon. Today, Goddard's 1920 paper is considered a classic in the field of rocketry. Copies of Goddard's paper reached Europe where it received more serious attention, especially in Germany. The German Rocket Society was formed in 1927, and in 1931 the German Army started a rocket program, culminating in the German V-2 rockets of World War II.

Robert Goddard was born in Wooster, Massachusetts, on October 5, 1882. When Goddard was a teenager, he read H. G. Wells's science fiction classic *War of the Worlds* and became obsessed with the idea of traveling to the moon or even Mars. As a student Worcester Polytechnic Institute and later at Clark University, Goddard filled notebook after notebook with equations and computations concerning rockets. His calculations convinced him that launching rockets into space would require fuels more powerful than the black powder that had been used in rockets since their invention by the Chinese. Liquid fuels like gasoline or liquid hydrogen mixed with liquid oxygen for combustion might do the trick. After earning a Ph.D. in physics and landing a job as an instructor at Clark University, Goddard was granted two U.S. patents in 1914: one was for the design of a multistage rocket and the other for a nozzled combustion chamber that allowed the use of a liquid fuel.

During the 1920s, Goddard performed experiments with liquid-fueled rockets on the farm of a distant relative, "Aunt" Effie Ward, in Auburn, Massachusetts. In 1923, he tested a rocket engine powered by gasoline and liquid oxygen; by the end of 1925, the engine could operate without the surrounding testing frame. The culmination of Goddard's work came on March 16, 1926, when he successfully launched the world's first liquid fuel rocket, a feat that rivals the Wright Brothers' first powered flight in its historical significance. The skinny, ten-foot-tall rocket reached an altitude of 41 feet, flew for 2.5 seconds, and landed 184 feet from the launching frame. Three years later, Goddard achieved another first at Auburn by launching a rocket carrying a scientific payload consisting of a camera and a thermometer; those instruments deployed successfully. This launch generated widespread publicity and brought Goddard's efforts to the attention of aviator Charles Lindbergh, who became an enthusiastic backer and fund-raiser.

In the 1930s, Goddard moved his base of operations to the desert outside of Roswell, New Mexico (yes, *that* Roswell). Here, he was the first to use a gyroscope to control the flight of a rocket and the first to use vanes in the rocket motor blast for guidance. In the late 1930s, Goddard tried unsuccessfully to interest the War Department in the possible military applications of rockets. The navy did, however, enlist his help in 1941 with rocket bombs and jet-assisted take-off, projects he worked on at the Naval Experiment Station in Annapolis, Maryland. Interestingly enough, Goddard had developed and demonstrated the basic principle of a "bazooka" back in 1918, just before the end of World War I. During World War II, one of Goddard's students, who had worked with him on the concept in 1918, continued the research that resulted in the bazooka. Goddard worked at the Naval Experiment Station until his death following a throat operation, on August 10, 1945, just four days after the first atomic bomb was dropped on Japan. Goddard's fundamental contributions to rocketry were not fully appreciated

Robert Goddard and his first liquid-fuel rocket launched from a Massachusetts farm in 1926.

until the appearance of the German V-2 rockets in 1943, weapons that utilized many of Goddard's ideas.

## Visiting Information

The Asa Ward farm where Goddard launched the first liquid fuel rocket has been converted into the Pakachoag Golf Course located on Upland Street in Auburn, Massachusetts. The sign for Upland Street is difficult to see—I missed it several times—so be sure to get clear directions before visiting. There is a parking lot across from the golf course. You will see a stone wall near the tee area. Simply follow the wall, and you will see a four-foot-tall granite obelisk almost directly ahead of you. The obelisk has the following inscription: "Site of launching of world's first liquid propellant rocket by Dr. Robert H. Goddard—16 March 1926." Along the street are two more markers in the corner of the golf course near some evergreen bushes (to your right as you face the golf course). One marker is a five-foot-tall granite tablet erected by the American Rocket Society in 1960. Part of the inscription reads "in recognition of this significant achievement in the evolution of astronautics." A final marker consists of a metal placard mounted on the side of a boulder; this National Park Service marker designates the site as a registered National Historic Landmark. Along Route 12 in Auburn next to a fire station is the Dr. Robert H. Goddard Park. The small park has a full-scale replica of Goddard's rocket along with a Navy Polaris Missile.

On the campus of Clark University in Worcester, Massachusetts, is the Robert H. Goddard Library with the Goddard Exhibition Room on the ground level. These displays include diaries and test reports, photographs, and rocket hardware. The exhibition is free and open to the public on weekdays from 9:30 A.M. To 4:00 P.M., but is closed on weekends and holidays. To make sure the exhibition is open, call the number at

Telephone: 508–793–7572
(Goddard Exhibition Room)

right. The Roswell Museum and Art Center's extensive exhibit on Goddard includes a re-creation of his workshop.

## Arthur Compton

Arthur Holly Compton was born into a distinguished academic and devoutly religious family in Wooster, Ohio, in 1892. His father was a Presbyterian minister and professor of philosophy at the College of Wooster where Arthur received his undergraduate degree. Compton's mother was a Mennonite who dedicated herself to missionary causes and was named American Mother of the Year in 1939. One of his two brothers became a diplomat and president of what is now Washington State University. His older brother Karl blazed a trail into physics and became president of the Massachusetts Institute of Technology. After graduating, the tall, handsome, and athletic Arthur considered a career in religion, but his father counseled him to pursue his interest in science, "Your work in this field may become a more valuable Christian service than if you were to enter the ministry or become a missionary." So in 1913, Arthur followed Karl to Princeton University to do graduate work in physics. In his early days at Princeton, Arthur devised an elegant new way of demonstrating the Earth's rotation. Then, Karl introduced Arthur to the study of X-rays, a phenomenon that became the focus of Arthur's scientific research for the next decade. Arthur, studying the angular distribution of X-rays reflected from crystals for his thesis, earned his Ph.D. in 1916. Upon graduation, he married Betty McCloskey, whom he had met at Wooster, and took a position as an instructor of physics at the University of Minnesota. The next year, he accepted a position as a research engineer at Westinghouse Corporation in Pittsburgh. In 1919, Compton was awarded a fellowship by the National Research Council, a program that provided many scientists of the 1920s and 1930s with an opportunity to freely explore their research interests. Compton used his fellowship to study at the Cavendish Laboratory at Cambridge University in England. The laboratory equipment was not well-suited for studying X-rays so he spent a year doing similar experiments on gamma rays. In 1920, he accepted an academic appointment as chairman of the physics department at Washington University in St. Louis, "a small kind of place," as Compton put it. There, out of the mainstream of physics, he perfected his apparatus and focused his work on X-rays, which eventually earned him a Nobel Prize.

In the early 1920s, physicists were still debating the nature of light. Was it a wave or a particle? Throughout the 1800s, evidence accumulated that seemed to point unambiguously toward a wave model of light. Then in 1905, Einstein resurrected the particle model to explain the photoelectric effect. According to Einstein, light consisted of little bundles or "quanta" of energy that later became known as photons. Compton put the particle model to the test by firing X-rays and gamma rays at electrons to see if the rays behaved like particles or waves. If the rays behaved like waves, then the scattered light would spread out in all directions, like water waves emanating from a rock dropped in a pool. If the rays behaved liked particles, then the interaction between the X-rays and electrons would be similar to the collision between billiard balls on a pool table. When Compton performed the experiment, he found that the energy of the X-rays was reduced after hitting the electrons. This decrease in energy showed up as an increase in wavelength, a shift that became known as the Compton Effect. Compton successfully explained his results by using the particle model of light along with the conservation of energy and momentum. Compton's experiment settled the question. When light interacts with matter, it behaves like a particle; photons are real. In 1923, Compton published a paper describing his results; he concluded, "This remarkable agreement between our formulas and the experiments can leave but little doubt that the scattering of X-rays is a quantum phenomenon." For this work, Compton was awarded a share of the 1927 Nobel Prize in physics. Today, the Compton Effect has applications in materials science and radiation therapy.

In 1923, Compton left Washington University to fill the professorship at the University of Chicago that had been vacated by Robert Millikan. During the 1930s, Compton led an extensive, worldwide study on how the intensity of cosmic rays varied across the surface of the Earth. The results clearly showed that cosmic ray intensity was linked to latitude as measured relative to the Earth's magnetic field. Because charged particles interact with a magnetic field, this finding proved conclusively that cosmic rays were charged particles streaming in from outer space.

During World War II, Compton played an important part in the Manhattan Project. He was put in charge of research on the production of plutonium, a chemical element that did not exist in nature. Plutonium couldn't be mined; it had to be made. The work was conducted at the University of Chicago at a facility that became known as the "Metallurgical Laboratory" or "Met Lab," a cover name that obscured the facility's real purpose: to produce a controlled chain reaction in a "pile" of uranium that would produce plutonium. In December 1942, Met Lab scientists under the direction of Enrico Fermi achieved a sustained chain reaction in the world's first nuclear reactor.

Toward the end of the war, Compton served on the Scientific Panel of the "Interim Committee" along with E. O. Lawrence, Enrico Fermi, and J. Robert Oppenheimer. This committee, established by President Harry Truman, considered possible uses for the atomic bomb against Japan. Compton believed that the bomb should be used because it would bring about a swift end to the war and thereby save countless American and Japanese lives.

Compton returned to Washington University in 1946 when he was named the university's chancellor. He resigned as chancellor in 1953 but continued as a faculty member. In 1961, he became a sort of professor-at-large and intended to divide his time between Washington University, the

University of California at Berkeley, and Wooster College. He died suddenly in 1962 at age sixty-nine at Berkeley and is buried in the Wooster Cemetery.

## Visiting Information

From the late 1920s until 1945 Compton lived in the house at 5637 South Woodlawn Avenue in Chicago's Hyde Park community. It has been designated as a National Historic Landmark and is a private residence. At Washington University, there is a plaque inside the main entrance of Eads Hall, commemorating the building where Compton discovered the X-ray scattering effect.

# Enrico Fermi

Enrico Fermi, born in Rome in 1901, was a central figure in twentieth-century physics, equally adept at both experimental and theoretical work, a combination rarely seen in a single physicist. His father was an administrator for the Italian railroad, and his mother taught elementary school. From an early age, Fermi displayed talent in science and mathematics. He also had mechanical skill and built motors and other electrical devices with his beloved older brother Guilio. When Fermi was fourteen, tragedy struck the family; Guilio died during minor throat surgery. Fermi dealt with his grief by immersing himself in science. He went to a market, found two antique Latin volumes on physics, bought them with his allowance, took them home and studied them intently. At age seventeen, Enrico took the entrance exam for the Scuola Normale Superiore, a special university for gifted students in Pisa. One section of the exam required him to write an essay on the theme "Characteristics of Sound." Enrico proceeded to derive a partial differential equation describing a vibrating rod and used Fourier analysis to solve it. After grading the test, the examiner called Fermi in, congratulated him on his essay, and predicted he would become an important scientist. Fermi earned his doctorate in physics, spent some time studying under Max Born in Göttingen and with Paul Ehrenfest in Leyden, and wound up as a lecturer at the University of Florence.

Fermi's first major contribution to physics came in 1926, when he discovered the statistical laws governing particles that obeyed the Pauli exclusion principle, a quantum mechanical law that places restrictions on the location of electrons in an atom. The laws became known as Fermi-Dirac statistics. In 1927, Fermi became a professor of physics at the University of

Rome, where, in the early 1930s, he developed a theory to explain some puzzling aspects of a type of radioactivity called beta decay (when an electron is spat out of the nucleus). Fermi postulated that beta decay was caused by a new fundamental force of nature, the "weak force." The success of Fermi's "weak force" in explaining beta decay convinced physicists to take the radical idea seriously. Experiments later confirmed the reality of the new force.

Fermi then began the work that earned him the Nobel Prize. Irene and Jean Joliot-Curie had found that bombarding nuclei with neutrons produced new kinds of radioactive atoms. Fermi took the Joliot-Curie discovery and ran with it, by creating new radioactive isotopes and discovering "slow neutrons." ( Nuclear reactors later used these slow neutrons.) In 1938, Fermi was awarded the Nobel Prize in physics for his "demonstrations of the existence of new radioactive elements produced by neutron irradiation, and for his related discovery of nuclear reactions brought about by slow neutrons." By that time, Benito Mussolini had begun an anti-Semitic campaign that threatened Fermi's Jewish wife, Laura, and their two children. The crackdown also put out of work most of Fermi's research team. When the Fermi family went to Stockholm to accept the Nobel Prize, instead of returning to Italy, they hopped on a ship and sailed to the United States. The Fermi's bought a house in Leonia, New Jersey, and Fermi accepted a position at Columbia University.

After hearing of the discovery of nuclear fission by two German scientists, a team at Columbia, including Fermi, conducted the first nuclear fission experiment in the United States. (See the entry on Columbia University in chapter 3.) Fermi commenced a series of experiments on nuclear fission and started commuting to the University of Chicago. With funding from the top-secret Manhattan Project, Fermi and his colleagues began building the Chicago Pile #1, the world's first nuclear reactor, beneath Stagg Field. (See the entry on the Nuclear Sculpture in chapter 7.) The reactor was successfully tested on December 2, 1942, and a coded message was telephoned to Washington: "The Italian navigator has landed in the New World." Fermi lived for a while at Los Alamos to serve as general advisor to the Manhattan Project. After the war in 1946, Fermi accepted a professorship at the Institute for Nuclear Studies at the University of Chicago, a position he held until his death, in 1954, from stomach cancer. When his colleague, Emilio Segré, visited Fermi in the hospital, he saw Fermi counting drops from his intravenous tube, timing them with a stopwatch to measure the flow rate.

## Visiting Information

Enrico Fermi lived in three houses during his time in the United States. In Leonia, New Jersey, the Fermi's lived at 382 Summit Avenue from 1940 through 1946. While working at Los Alamos, he lived at 1300 20th Street. Finally, in Chicago, they lived at 5537 South Woodlawn Avenue in the Hyde Park neighborhood from 1946 until Enrico's death in 1954. This final house has a "Chicago Tribute" marker in front. None of the houses is open to the public. Enrico Fermi is buried at the Oak Woods Cemetery in Chicago.

# J. Robert Oppenheimer

Often referred to as the father of the atomic bomb, J. Robert Oppenheimer was the scientist who led the Manhattan Project, the effort to build an atomic bomb during World War II. Oppenheimer, who was six-feet tall but never weighed more than 125 pounds, was selected for the job by General Leslie Groves in spite of opposition from army intelligence officers who balked at Oppenheimer's left-wing background. Indeed, Oppenheimer's wife, brother, sister-in-law, and former fiancée had, for a time, been members of the Communist Party. Besides, Oppenheimer was a theoretical physicist, and most of the project's work was experimental. Another strike against Oppenheimer was the fact that he had never won a Nobel Prize, a recognition many other project leaders had received. Nevertheless, Groves, impressed with Oppenheimer's expansive knowledge and off-the-scale intellect, pushed the appointment through. It proved to be an inspired choice. Noted science historian Gerald Holton writes: "It is generally agreed that no one could have directed so well the large group of prima donna scientists assembled at Los Alamos under the difficult and panic-evoking condition of war."

Robert Oppenheimer was born on April 22, 1904, into a wealthy nonpracticing Jewish family who lived in a spacious New York City apartment. His father was a successful textile importer, and his mother, a teacher and painter. Gifted, but frail and frequently ill, Oppenheimer attended the private Ethical Culture School, an extension of Columbia University educator Felix Adler's Society for Ethical Culture movement, based on the idea that man himself, not God, was responsible for his life and his destiny. His interest in science began at an early age; a professional-grade microscope was a favorite toy, and collecting rocks and minerals ranked as a favorite hobby. In the third grade he performed experiments, in the fourth, he began keeping a scientific notebook, and in the fifth, he began studying physics. At age twelve, he gave a lecture to the New York Mineralogical Club, whose stunned membership had assumed, based on the maturity of his letters, that he was an adult.

Oppenheimer graduated as valedictorian of his class in 1921 and entered Harvard University as chemistry major. During a course in thermodynamics, taught by future Nobel laureate Percy Bridgeman, Oppenheimer felt himself drawn instead to physics. After graduating summa cum laude in only three years, Oppenheimer studied physics at the Cavendish Laboratory at Cambridge University. During his time in England, the most troubled period of his life, he became emotionally unstable. In one incident, he tried to strangle a good friend. In another, he dosed an apple with poisonous chemicals from his lab and put it on the desk of his tutor; it was fortunate that the apple was left uneaten. When university administrators discovered what had happened, however, Oppenheimer was put on probation and ordered to see a psychiatrist. Part of the problem was that Oppenheimer hadn't found his niche in physics. At the lab, he was spending his days laboriously making thin films of beryllium for an experiment he never completed. He was a klutz in the lab and confessed to a friend that he couldn't even solder two copper wires together. Oppenheimer discovered that his talents were far more suited for theoretical work, and in 1926 he transferred to what at the time was the world's leading center of theoretical physics, the University of Göttingen in Germany. Here, he learned about the new theory of quantum mechanics from many physicists who had made original contributions to the development of the theory, and he worked especially closely with Max Born. Between 1926 and 1929, Oppenheimer published sixteen scientific papers, which earned him an international reputation as a theoretical physicist. After interrogating Oppenheimer during his oral exam for his Ph.D., a professor supposedly said, "Whew, I'm glad that's over. He was on the point of questioning me!"

Back in the United States, Oppenheimer accepted a joint appointment at Berkeley and Caltech. For the next thirteen years, he split his time between the two schools spending the fall semester at Berkeley and the spring term at Caltech. Oppenheimer was a superb teacher, mentoring a new generation of physicists, and is often credited as one of the founding fathers of the American school of theoretical physics. During the 1930s, Oppenheimer made important contributions to the theory of the positron, cosmic ray showers, and quantum tunneling. With a colleague, Oppenheimer wrote papers showing there was an upper limit to the mass of a stable neutron star.

A star above this limit would undergo further gravitational collapse. The papers hinted at the existence of what we now call black holes. While at Caltech, Oppenheimer became close friends with chemist Linus Pauling. Together, they planned to tackle the problem of the nature of the chemical bond. But their personal and professional relationship ended abruptly when Pauling learned from his wife that Oppenheimer had invited her to join him on a private trip to Mexico.

Oppenheimer, named the scientific director of the Manhattan Project in 1942, presided over what has to be the greatest concentration of scientific genius the world has ever seen. Early on the morning of July 16, 1945, the world's first nuclear weapon, tested in the desert of New Mexico, prompted the now-famous reaction from Oppenheimer: "Now I am become death, the destroyer of worlds." After the successful test, Oppenheimer served on the committee that recommended using the bomb against the Japanese, a decision he later regretted.

After the war, Oppenheimer, who was now a nationally known figure, was appointed director of the Institute for Advanced Study in Princeton, a position he held for the rest of his life. He also served as chairman of the General Advisory Committee of the Atomic Energy Commission (AEC), a civilian agency responsible for formulating nuclear policy. As chairman, Oppenheimer argued for international arms control and research into the peaceful applications of nuclear energy and against an escalating arms race with the Soviet Union. When the question of building the hydrogen bomb was raised, Oppenheimer initially recommended against it—partly on moral grounds and partly for a technical reason: he didn't think the H-bomb had much chance of working. He later changed his mind when a feasible design was proposed. Oppenheimer's opinions did not set well with proponents of a nuclear arms build-up. This, in addition to his occasional sarcasm and contemptuous attitude, earned him plenty of enemies. In particular, AEC commissioner Lewis Strauss held a grudge against Oppenheimer because he had embarrassed Strauss at a congressional hearing.

Strauss's campaign against Oppenheimer reached fruition when President Dwight Eisenhower, presented with new evidence of Oppenheimer's supposed communist sympathies, asked Oppenheimer to resign from the committee. Oppenheimer refused, requesting a public hearing so that he could defend himself against the accusations. Meanwhile, his security clearance was suspended. At the hearing, a number of scientists came to the defense of Oppenheimer with one important exception: Edward Teller, the

Robert Oppenheimer and General Leslie Groves inspecting debris at Trinity Site, July 1945.

father of the H-bomb. In his testimony, Teller did not question Oppenheimer's loyalty, but rather his judgment. When asked if it would endanger the common defense and security to grant clearance to Oppenheimer, Teller said that "Dr. Oppenheimer's character is such that he would not knowingly and willingly do anything designed to endanger the safety of this country. . . . If it is a question of wisdom and judgment, as demonstrated by judgment since 1945, then I would say one would be wiser not to grant clearance." In the paranoid atmosphere of the McCarthy era, Oppenheimer could not quell the suspicions raised by the friends that he kept, and Teller's testimony didn't help his cause. His security clearance was permanently revoked, and he was pilloried by the press. Teller was ostracized by a scientific community angered by Oppenheimer's treatment.

During the last years of his life, Oppenheimer lectured and wrote about the role of science in society and about intellectual ethics and morality. In 1963, President Lyndon Johnson, in a gesture signifying a sort of political rehabilitation, presented Oppenheimer with the Enrico Fermi Award "for contributions to theoretical physics as a teacher and originator of ideas, and for leadership of the Los Alamos Laboratory and the atomic energy program during critical years." A chain smoker for his entire adult life, Oppenheimer died of throat cancer in 1967 at his home in Princeton. He was cremated, and his ashes were dropped into the sea near his beach house on St. John, in the Virgin Islands.

### Visiting Information

The Manhattan apartment that was Robert Oppenheimer's childhood home is located at 155 Riverside Drive at West 88th Street. The family lived on the eleventh floor overlooking the Hudson River. The apartment was decorated with original artwork by Picasso, Rembrandt, Renoir, and Van Gogh. While at Berkeley, he lived in a Spanish-style house at #1 Eagle Hill at the end of a cul-de-sac in Kensington, a short drive from campus. He also lived for a time at 10 Kenilworth Court. During the Los Alamos years, he lived at the house at 1967 Peach Street. His vacation home on St. John is situated on the northern section of beautiful Gibney Beach, an area that the locals refer to as "Oppenheimer Beach." After Oppenheimer's daughter died, she left the property to "the people of St. John for a public park and recreation area." Today, the property has been converted into a Community Center.

## Richard Feynman

Richard Feynman, a bongo-drum-playing, hallucinogenic-drug-taking, topless-bar-frequenting, physicist extraordinaire, has a cult following among the alpha-nerd crowd second only to Albert Einstein himself. The colorful and iconoclastic Richard Phillips Feynman was born in 1918 in Far Rockaway, in the borough of Queens, on the fringe of New York City. His father was a salesman who encouraged his son to be curious about the natural world around him and taught young Richard that merely knowing the name of something doesn't mean that you understand it. From his mother, Richard acquired a talent for storytelling and a playful sense of humor. The family was Jewish and attended synagogue, but they were neither rigid nor ritualistic in their religious practices. As an adult, Feynman was an avowed atheist, a belief he shared with many other twentieth-century physicists. Feynman didn't start speaking until age two and had a measured IQ of 125, high for the general population, but rather modest for a theoretical physicist.

When he was eleven or twelve, Feynman converted a refrigerator-sized wooden box into a laboratory and placed it in the corner of his bedroom. He experimented with light bulbs, switches, and fuses, devised a burglar alarm for his room, and burned holes in paper with a spark from a coil. Later, he was fascinated with radios, exploring the mysteries of their electrical entrails, repairing broken sets, and listening to broadcasts with earphones as he fell asleep. He attended Far Rockaway High School where he impressed his classmates by solving a myriad of mind and math puzzles. He mastered calculus by age fifteen and, as a senior, won the New York University Math Championship. When he applied to Columbia University, he was rejected because of a strict quota capping the number of Jewish students. He went to MIT instead, where he took every physics course offered, thereby neglecting his studies in subjects outside the scientific and mathematical arena. When he was a senior, he was invited to compete in a prestigious national mathematics competition, consisting of a dozen or so extremely difficult mathematical problems. In some years, the median score was zero, meaning that half the competitors were unable to solve a single problem correctly. Feynman won the competition; the large gap between Feynman's score and the second-place finisher shocked the judges. After graduating in 1939, Feynman went to Princeton, where he had earned an unprecedented perfect score on the math and physics entrance exams. Feynman worked under the imminent physicist John Archibald Wheeler, became a leading expert on quantum mechanics, and earned his doctorate with his dissertation, "The Principle of Least Action in Quantum Mechanics."

While still a student at Princeton, Feynman was recruited to work on the Manhattan Project and moved to Los Alamos in 1943. There, he worked in the theoretical division under Hans Bethe and derived the Bethe-Feynman formula for calculating the yield of a fission bomb. Feynman's main responsibility was overseeing a group who performed the tedious mathematical computations required to convert theory into reality. For amusement, Feynman occupied himself by toying with the lab's security safeguards, picking locks, and cracking safes.

Against his family's wishes, Feynman had married Arline Greenbaum who had been diagnosed with tuberculosis, a contagious and, at that time, fatal disease. The couple took extraordinary precautions, and Feynman never contracted the disease. While he was at Los Alamos, she stayed at a hospital in Albuquerque, where Richard visited her nearly every weekend. Arline succumbed to the disease in 1945. Their love affair is chronicled in the 1996 movie *Infinity*.

After the war, Feynman followed Hans Bethe, whom he liked and admired, to Cornell University where Feynman taught theoretical physics. He was feeling a little burned out with physics and was not making much progress on anything until one day, while sitting in the cafeteria, a student tossed a plate into the air. Feynman noticed that the plate was wobbling and observed that the red Cornell medallion decorating the plate was rotating faster than the wobbling motion. Just for fun, Feynman decided to derive the equation of motion for the plate. Although the wobbling plate problem was of no real importance, the incident rescued Feynman from his rut and set him onto the path that lead to the Nobel Prize.

In 1947, Feynman refined the theory of Quantum Electrodynamics (QED), a quantum field theory of the electromagnetic force whose main characters are electrons and photons. In Feynman's version of QED, two electrons exert a repulsive electrical force on each other by exchanging a "package" (or "quantum") in the form of a photon. Feynman invented a schematic way of representing these interactions. These "Feynman Diagrams" continue to be a useful tool in today's string theory and M-theory. For this work, Feynman shared the 1965 Nobel Prize with Julian Schwinger and Sin-Ituro Tomonaga, who developed versions of QED independent from Feynman. When Feynman bought a Dodge Tradesman Maxivan in 1975, he proudly decorated it with Feynman diagrams.

Feynman grew weary of the winter weather at Cornell and complained of having to put chains on his tires so, in 1951, he accepted a position at Caltech. There, his major scientific accomplishments included a quantum mechanical explanation of the weird, gravity-defying behavior of liquid helium at very low temperatures, a phenomenon known as superfluidity. This work later served as a foundation for understanding superconductivity,

Richard Feynman standing in his Caltech office.

although the explanation eluded Feynman. Working with Murray Gell-Mann, Feynman also developed a general theory of the weak interaction.

Feynman was not only a great physicist but also an excellent teacher with a knack for explaining complicated ideas in an understandable way. In 1963, Feynman was asked to teach the introductory physics classes at Caltech. The goal was to introduce modern topics into the freshman and sophomore physics curricula so that the student's fervor for physics would not be bludgeoned out of them with too many pulleys, levers, and inclined planes. Feynman's lectures were successful, but not with the intended audience. As the weeks went by, attendance by the undergraduates went down while attendance by graduate students and faculty went up. The highly original lectures were later published as *The Feynman Lectures in Physics* and have sold more than three million copies around the world. Nearly every serious student of physics has a copy of this classic on his or her bookshelf.

In 1986, Feynman was asked to serve on a panel to investigate the explosion of the space shuttle *Challenger*. Acting on hints dropped by another panel member, Feynman got hold of an O-ring, one of the circular rubber seals that separated sections of the solid rocket boosters. Feynman tested the O-ring and found that the material lost some of its flexibility at temperatures comparable to those on the morning of the ill-fated launch. This reduction in resiliency allowed super-heated gas to escape from the booster, causing an explosion. In a dramatic moment captured on television, after a NASA manager had repeatedly testified that the O-rings would retain their flexibility even in extreme cold, Feynman submerged a piece of O-ring material that was clamped flat into a glass of ice water. When he removed the clamp, the material remained in its flattened position, clearly demonstrating its loss of resiliency. Feynman, in a harsh criticism of NASA that was included in an appendix to the final report, accused NASA of "playing Russian roulette" with the lives of astronauts. Feynman's experience on the *Challenger* commission is chronicled in the book *What Do You Care What Other People Think?*

Feynman was a quirky and fascinating character. He had an aversion to music and yet played the bongo drums, an instrument he had learned during a sabbatical year spent in Brazil. He had liberal views on sexuality and gave advice on picking up women in his autobiography *Surely You're Joking, Mr. Feynman*. While at Caltech, he used a topless bar as a second office. He experimented with LSD and marijuana and a drug called ketamine, which he used in conjunction with a sensory deprivation tank to induce hallucinations. Late in his life, he vowed to travel to Tannu Tuva, a remote location in Russia that his father had told him about when he was a child. Permission to visit was slow in coming due to the bureaucracy of the former Soviet Union. Permission was finally granted in 1988, but it came too late, just before Feynman died of cancer. Some friends of Feynman made the journey for him, and a plaque to the memory of Richard Feynman can be found at the Center of Asia Monument in Kyzyl.

### Visiting Information

Richard Feynman's boyhood home is located at 792 Richard Feynman Way in Far Rockaway, Queens. The street, formerly named Cornaga Avenue, and before that, New Broadway, was renamed to honor Feynman. The house, a private residence, is not open to the public. While at Los Alamos, Feynman lived in a bachelor dormitory that is now a Unitarian Church at the northwestern corner of Sage Loop and 15th Street. Feynman is buried in the Mountain View Cemetery in Altadena, California.

| 4 Be Beryllium 9.0122 |
| 12 Mg Magnesium 24.305 |

| 20 Ca Calcium 40.078 | 21 Sc Scandium 44.9559 | 22 Ti Titanium 47.867 | 23 V Vanadium 50.9415 | 24 Cr Chromium 51.9961 | 25 Mn Manganese 54.938 | 26 Fe Iron 55.845 | 27 Co Cobalt 58.9332 | 28 58.6934 | 63.546 | 65.409 | 69.723 | 72.64 | 74.9216 | 78.96 | 79.904 | 83.79 |
| 38 Sr Strontium 87.62 | 39 Y Yttrium 88.9059 | 40 Zr Zirconium 91.224 | 41 Nb Niobium 92.9064 | 42 Mo Molybdenum 95.94 | 43 Tc Technetium (98) | 44 Ru Ruthenium 101.07 | 45 Rh Rhodium 102.9055 | 46 Pd Palladium 106.42 | 47 Ag Silver 107.8682 | 48 Cd Cadmium 112.411 | 49 In Indium 114.818 | 50 Sn Tin 118.71 | 51 Sb Antimony 121.76 | 52 Te Tellurium 127.6 | 53 I Iodine 126.9045 | 54 Xe Xenon 131.29 |
| 56 Ba Barium 137.327 | 72 Hf Hafnium 178.49 | 73 Ta Tantalum 180.9479 | 74 W Tungsten 183.84 | 75 Re Rhenium 186.207 | 76 Os Osmium 190.23 | 77 Ir Iridium 192.217 | 78 Pt Platinum 195.078 | 79 Au Gold 196.9665 | 80 Hg Mercury 200.59 | 81 Tl Thallium 204.3833 | 82 Pb Lead 207.2 | 83 Bi Bismuth 208.9804 | 84 Po Polonium (209) | 85 At Astatine (210) | 86 Rn Radon (222) |
| 88 Ra Radium (226) | 104 Rf Rutherfordium (261) | 105 Db Dubnium (262) | 106 Sg Seaborgium (266) | 107 Bh Bohrium (264) | 108 Hs Hassium (277) | 109 Mt Meitnerium (268) | 110 Ds Darmstadtium (271) | 111 Rg Roentgenium (272) | 112 Uub Ununbium (277) |

| 57 La Lanthanum 138.9055 | 58 Ce Cerium 140.116 | 59 Pr Praseodymium 140.9077 | 60 Nd Neodymium 144.24 | 61 Pm Promethium (145) | 62 Sm Samarium 150.36 | 63 Eu Europium 151.964 | 64 Gd Gadolinium 157.25 | 65 Tb Terbium 158.9253 | 66 Dy Dysprosium 162.5 | 67 Ho Holmium 164.9303 | 68 Er Erbium 167.259 | 69 Tm Thulium 168.9342 | 70 Yb Ytterbium 173.04 | 71 Lu Lutetium 174.96 |
| 89 Ac | 90 Th | 91 Pa | 92 U | 93 Np | 94 Pu | 95 Am | 96 Cm | 97 Bk | 98 Cf | 99 Es | 100 Fm | 101 Md | 102 No | 103 Lr |

# 2
# Chemists

*The chemists are a strange class of mortals, impelled by an almost*
*insane impulse to seek their pleasures amid smoke and vapour,*
*soot and flame, poisons and poverty; yet among all these evils*
*I seem to live so sweetly that may I die if I were to change places*
*with the Persian king.*

Johann Joachim Becher

Everything, be it natural or synthetic, is made of chemicals. Chemists study chemicals to determine the composition, structure, and properties of matter and to derive the laws that govern how substances combine and react with each other. Some chemists apply the knowledge gained from basic chemical research to develop new products and processes or to improve existing ones. Applied chemical research has resulted in thousands of practical products that improve the quality of our lives—from cleaners and cosmetics to paints and plastics. We use chemicals to look good, smell good, clothe our bodies, clean our homes, run our cars, grow and preserve our food, and prevent and treat diseases. In addition to beakers, test tubes, and other assorted glassware, chemists use sophisticated instrumentation to collect and analyze data and simulate phenomena. Computers enable chemists to perform combinatorial chemistry, a technique whereby huge numbers of compounds can be tested simultaneously to uncover those with desirable properties.

The science of chemistry has its roots in the use of fire to purify metals and to form alloys. When gold became a precious commodity, people began trying to transform other more common and lowly substances like lead into gold. This marked the beginning of the pseudoscience of alchemy. Chemistry finally became distinct from alchemy when Robert Boyle published *The Sceptical Chymist* in 1661. Still, for the next hundred years or so, chemists continued searching for imaginary substances that, as it turned out, simply didn't exist; such substances included the élan vital, the essence of life that transformed inanimate substances into living organisms, or phlogiston, the material that made things burn. Frenchman Antoine Lavoisier, during the latter half of the eighteenth century, finally brought chemistry into the modern age. Lavoisier had at his disposal the finest private laboratory in the world, and he used it to bring rigor and methodology to chemistry. He disposed of the unproductive idea of phlogiston and identified oxygen. Most important, he discovered the concept known as the conservation of mass, the idea that matter can change forms but not vanish. Chemistry took a giant step in the direction of orderliness and clarity with the work of a wild-and-crazy-looking Russian chemist named Dmitri Mendeleev. Inspired by the card game we know as solitaire, Mendeleev's insight was to arrange the elements in a table with the horizontal rows in order of increasing atomic number and the vertical columns consisted of elements with similar properties. The resulting periodic table has been rightly called the most elegant organizational chart in history.

Today's chemists, like physicists, are specialists. Analytical chemists analyze samples of material to figure out their structure and composition, a task critical to the pharmaceutical industry. Organic chemists study carbon compounds that form the basis for all life on Earth. Inorganic chemists study compounds that do not contain carbon, such as those used in electronics. Physical chemists study the physical properties of substances including the dynamics of chemical reactions and the flow and storage of energy in a reaction. Other areas of specialization include theoretical chemistry, materials chemistry, and medicinal chemistry. There are about 85,000 career chemists in the United States employed outside academia; about 41 percent work for manufacturing firms mainly in the chemical manufacturing industry, another 18 percent in scientific research and development, and 12 percent in architectural, engineering, and related services. Another 25,000 chemists are employed by colleges and universities.

Each entry below provides a biographical sketch of a celebrated chemist along with a place or places you can visit to make a connection to these scientific luminaries. The entries are arranged in chronological order, according to the chemist's birth date.

## Joseph Priestley

Joseph Priestley, the chemist who discovered oxygen, was born in Fieldhead, England, in 1733. He and his family were Dissenters, a label used for people who did not conform to the beliefs of the Church of England. At age nine, he went to live with a wealthy and childless aunt and uncle. His aunt recognized young Joseph's intelligence and sent him to the local schools where he learned Greek, Latin, and Hebrew. At age eleven, he performed his first experiment—an investigation into how long spiders could live in bottles without fresh air. In his teens, he was tutored by a clergyman who introduced him to mathematics, science, and logic. His educational preparation was good enough for entry into England's great universities at Oxford and Cambridge, but, as a Dissenter, he was barred from attending. He enrolled instead at the Daventry Academy, a celebrated school for Dissenters. After his third year at Daventry, he decided to dedicate himself to the ministry, describing it as "the noblest of all professions." After graduating, Priestley lived and worked in a number of English towns and earned a living by preaching, teaching, and writing. Although he claimed that science was only a hobby, it was one he loved and took seriously. Priestley took a practical rather than theoretical approach to science, modeling himself after his friend and mentor Benjamin Franklin, whom he had met on a trip to London. With Franklin's encouragement, Priestley performed research in the area of electricity, which culminated in *The History and Present State of Electricity.* The book, coupled with his growing reputation as an experimentalist, earned him the distinction of being named a fellow of the Royal Society in 1766.

While drawing diagrams for a second book on electricity intended for the general public, Priestley discovered that rubber from India would cleanly erase pencil lines; thus, he invented that most indispensable of writing tools, the eraser. Priestley next discovered a convenient method for making soda water. Although Priestley did not take advantage of the commercial potential of his breakthrough, others, most notably J. J. Schweppe, did. The next year, the Earl of Shelbourne, after some prompting by Franklin, invited

Priestley to come to his estate to tutor his children, serve as his librarian, and act as his personal assistant and advisor. The position provided Priestley with financial security and gave him the free time he needed to carry out the research that would eventually earn him a place in scientific history.

The ancient Greeks identified earth, air, fire, and water as the four basic building blocks of the universe. This idea remained unchallenged until Priestley's time. Priestley himself wrote that few concepts "have laid firmer hold upon the mind" than the idea that air "is a simple elementary substance, indestructible and unalterable." In a series of experiments, Priestley discovered that air is not an elementary substance after all, but rather, a mixture of different gases. His most famous experiment took place on August 1, 1774, when he used a lens to focus sunlight on a sample of mercuric oxide. He isolated a gas "five or six times as good as common air" that caused a flame to burn intensely and allowed a mouse to stay alive much longer than the same amount of regular air. At the time, scientists incorrectly believed that flammable materials contained an essence called phlogiston that made them burn. Priestley called the gas "dephlogisticated air" because he thought it had no phlogiston in it and could therefore absorb the maximum amount. This explained why it burned so intensely. The French chemist Antoine Lavousier later named the gas "oxygen." Priestley's discovery proved that the ancient Greek model of four elements was wrong and opened the way to a revolution in chemistry based on the ideas, advanced by Lavousier, of chemical elements and compounds. Priestley, however, clung to his own revised version of phlogiston theory and rejected the new chemistry.

For reasons that are not completely clear, Priestley had a falling out with Shelbourne around the year 1779 and moved to take a ministerial position in Birmingham. There, Priestley was invited to join the Lunar Society, an illustrious group of manufacturers, inventors, and scientists that included engineers James Watt and Matthew Boulton, potter Josiah Wedgewood, and Erasmus Darwin, grandfather of Charles Darwin. The group met on nights of the full moon so that the moonlight could help the members find their way home.

During his Birmingham years, Priestley's writings on theological issues became more radical and iconoclastic. Priestley helped found the Unitarian Church and, in his writings, rejected the idea of the Trinity and asserted that the Protestant Reformation had not really reformed anything. In Priestley's own words, "We are . . . laying gunpowder . . . under the old building of

Joseph Priestley's house in Northumberland, Pennsylvania.

error and superstition . . . so as to produce an instantaneous explosion." Priestley's inflammatory language earned him the nickname "Gunpowder Joe." Priestley further argued that religion could not form the basis for a civil society and should be limited to private life. Priestley further thumbed his nose at authority by supporting both the American and French revolutions.

Public outrage against the religious Dissenters and supporters of revolutions boiled over in the Birmingham riots of 1791. An angry, drunken mob took a torch to Priestley's house and laboratory. The family lost everything and barely escaped with their lives. They fled to London, but things kept getting worse. Priestley was burned in effigy alongside Thomas Paine, received hate mail comparing him to Guy Fawkes and the devil, and was ostracized by long-time friends. When penalties against those who spoke out against the government became more severe, Priestley decided it was time to go. In 1794, he left England and came to America. Just five weeks after Priestley left, the British government began arresting radicals for treason.

Upon arriving in the United States, Priestley turned down an offer to teach chemistry at the University of Pennsylvania and settled instead in the remote hamlet of Northumberland to be near his sons who had immigrated to America a year earlier. Priestley continued to do scientific research and, in

1799, succeeded in isolating the poisonous gas carbon monoxide. Although his scientific productivity was waning, his mere presence stimulated interest in chemistry in the United States. Priestley also helped found the First Unitarian Church of Philadelphia, establishing a presence for the Unitarian church in America. Although he had hoped to avoid politics in his new country, political controversies continued to plague him until his good friend, Thomas Jefferson, was elected president in 1800. On February 3, 1804, Priestley began his final experiment but soon found himself too weak to continue and retired to his bed in the library. Three days later, he died there.

## Visiting Information

Joseph Priestley lived in a Georgian-style house on a hill overlooking the Susquehanna River in Northumberland, Pennsylvania, from 1798 until his death in 1804. The house and grounds were intended to re-create an English gentleman's estate. Joseph's youngest son Henry died in 1796, at age eighteen, probably from malaria, and his wife Mary had died a year later, before the house was finished. Joseph invited his oldest son, Joseph Jr., and his wife and five children, to come live with him in the new house. Priestley often held church services and sometimes taught school in the home. His library held more than sixteen hundred books. After Priestley's death, his son's family continued to live in the house until 1811 when he and his family returned to England. In 1874, chemists from around the country met at this house to commemorate the centennial of Priestley's discovery of oxygen. The meeting inspired the formation of the American Chemical Society. Today, the house is furnished with objects representative of the time period and the laboratory contains exact reproductions of his equipment. The house is located at 472 Priestley Avenue in Northumberland. Your visit begins in the Visitor Center where you can read some information panels and view a ten-minute film on the life of Joseph Priestley. You can also purchase tickets for the guided tours that take you through the rooms of the house. Prices are $4 for adults and $2 for children. The house is open Wednesday through Saturday from 9:00 A.M. until 4:30 P.M. and on Sundays from noon until 4:30 P.M. Tours leave on

> Website: www.josephpriestleyhouse.org
> Telephone: 570–473–9474

the hour starting at 10:00 A.M. There's not that much to see in the Visitor Center so try to time your visit so that you arrive close to the hour. It may be a good idea to call and let them know you are coming. The Priestley House is closed in January, February, and the first part of March. The Joseph Priestley grave site is only a few blocks away as is the Joseph Priestley Memorial Chapel. Maps to these sites are available at the Priestley House. The grave is marked by a taller, modern gravestone that was recently placed there by the American Chemical Society. The small, original marker is hidden behind the larger stone.

## George Washington Carver

George Washington Carver was born into slavery probably in spring 1865 in Diamond Grove, Missouri. (Slavery continued in Missouri until the state adopted a new constitution on July 4, 1865.) His mother, Mary, was owned by Moses Carver, a farmer who was philosophically opposed to slavery but was also pragmatic and needed help on his 240-acre farm. Carver never knew his parents. His father was killed in an accident on a neighboring farm before George was born. While George was a baby, he, along with his sister and mother were kidnapped by Confederate bandits. George was later rescued, but his mother disappeared forever. Moses and his wife Susan raised George and his brother Jim as their own. Due to his frail condition, possibly caused by a respiratory disease, George avoided work in the fields and instead helped Susan with a variety of domestic chores including gardening. As a result, George developed an early interest in plants.

At age eleven, Carver left the farm to attend a school for blacks in Neosho, the county seat. Carver didn't learn much from his poorly trained teacher, but he benefited greatly from his friendship with Mariah Watkins, in whose house he stayed. Watkins shared with Carver her knowledge of medicinal plants along with her deep religious faith. After Neosho, Carver wandered Missouri and Kansas for more than a decade. He attended high school, tried homesteading, and earned money by using the domestic skills he had learned. He applied to college, was accepted, and then turned away when he showed up on campus and administrators saw that he was a black man.

In the late 1880s, Carver found himself in Winterset, Iowa, where he met the Milhollands, a white couple who took an interest in Carver and encouraged him to apply to nearby Simpson College, a small, progressive, Methodist school that welcomed all qualified applicants regardless of race or gender. Carver applied, was accepted, and later wrote of the school: "They made me believe I was a real human being."

Carver studied art at Simpson, but his teacher, Etta Budd, didn't think Carver would be able to make a living as an artist. Budd just happened to be the daughter of the chairman of the horticulture department at Iowa State University in Ames, and, when she learned of Carver's interest in plants, she encouraged him to transfer. He did and stayed until he had earned a master's degree in agriculture, becoming the only African American in the country with graduate training in "scientific agriculture."

He received several attractive job offers, but the African American educator Booker T. Washington persuaded Carver to join him at the Tuskegee Institute in Alabama. Unfortunately, Carver and Washington did not get along with each other very well. Carver was a disheveled, disorganized dreamer, while Washington was a tidy, well-organized pragmatist. Carver wanted to concentrate on his research while Washington expected Carver to manage the school's farms, teach a full load of classes, serve on committees, sit on the executive council, and oversee the institute's sanitary facilities in addition doing research. Carver threatened to resign on a number of occasions, and Washington reacted by mollifying Carver because he recognized that he was a gifted researcher. After Washington's death in 1915, his successor demanded less of Carver and even relieved him of his teaching duties, except during the summer. Carver stayed at Tuskegee for the rest of his life.

Carver wanted to help impoverished southern farmers who were trying to grow crops in poor quality soil that had been used to grow cotton, a plant that depletes soil of nutrients. Carver showed that the soil could be enriched by planting legumes, such as peas and beans. It turns out that the root nodules of legumes take chemically inert molecular nitrogen from the air and transform it into nitrogen compounds that are essential for life, a process known as nitrogen fixation. Plants use nitrogen compounds to form chlorophyll molecules, which are used in photosynthesis. On his ten-acre experimental farm at Tuskegee, Carver dramatically increased crop yields by planting legumes and using good cultivation practices. For example, on one half-acre plot, he increased the output of sweet potatoes from 40 bushels to 266 bushels in only a few years. Carver wrote and distributed a series of free, easy-to-read bulletins that explained his recommended cultivation techniques to farmers.

But to convince farmers to give up lucrative crops like cotton and tobacco, Carver had to show a ready market for soybeans, peas, and sweet potatoes. He and his assistants went into the laboratory to develop new uses for these crops, an effort that later became formally known as the

George Washington Carver in the laboratory circa late 1930s.

"chemurgy" movement. Over the years, Carver created hundreds of new uses for a variety of crops, but Carver's virtuosity with the peanut made him famous.

Carver liked peanuts because they enriched the soil, were easy to cultivate, and were a good source of protein for poor farmers who couldn't afford much meat. Carver developed more than three hundred new products from peanuts including milk, cooking oil, various punches, medicinal products, and cosmetics. Still, Carver was not widely known until 1921, when he was called to testify before the House Ways and Means Committee as an expert witness on behalf of the peanut industry. American peanut farmers, being undercut by imported peanuts from China, were seeking tariff protection. At the hearing, Carver carefully laid out his samples of peanut-based products on the table and began describing them. His presentation was scheduled for ten minutes, but the spellbound committee repeatedly extended his time. Upon finishing, he received a standing ovation. The peanut industry got its tariff, and Carver became nationally famous as the "peanut man."

More important than Carver's scientific work, the significance of which has perhaps been exaggerated, was the fact that he stood as a proud symbol of African American achievement. During the 1920s, several progressive southern newspapers praised Carver for his accomplishments and used him as an example to promote improved racial relations and economic opportunity for African Americans. In 1941, *Time* magazine dubbed him as the "Black Leonardo."

Carver died in 1943 as a result of complications from a fall down a flight of stairs. He was buried at the Tuskegee Institute ironically next to his former nemesis, Booker T. Washington. His gravestone reads: "He could have added fortune to fame, but caring for neither, he found happiness and honor in being helpful to the world." A movement to establish a national monument in honor of Carver had begun even before his death. In July 1943, President Franklin D. Roosevelt designated $30,000 to establish near Diamond, Missouri, where Carver had spent his childhood, a national monument, the first dedicated to an African American and the first to honor someone other than a president.

## Visiting Information

Visitors will find several things to see and do at the George Washington Carver National Monument. At the Carver Science Lab, you can catch a park ranger performing various demonstrations, including one showing how to make peanut paper. The Carver Museum houses exhibits, dioramas, and artifacts that reveal what life was like on the farm. But the main attraction is the three-quarter-mile Carver Nature Trail that leads you past several points of interest, beginning with a statue depicting Carver as a young boy. He is shown sitting on a rock holding a plant, roots and all, in his left hand. The next stop is the spot where Carver was born. The original 12-by-12-foot cabin was destroyed by a tornado, but the place where the cabin once stood is outlined by wooden timbers. Go inside the 1881 Moses Carver House where period furniture and artifacts are on display. Several Carver family members, including Susan and Moses, are buried at the cemetery, which dates to 1835. The Carver National Monument is located near the town of Diamond, in southwestern Missouri. It is open daily, except for major holidays, from 9:00 A.M. until 5:00 P.M. There is no admission charge. Carver fans may

| Websites: | George Washington Carver National Memorial: www.nps.gov/gwca Tuskegee Institute National Historic Site: www.nps.gov/tuin |
| Telephone: | 417–325–4151 (George Washington Carver National Memorial) 334–727–3200 (Tuskegee Institute National Historic Site) |

also want to stop by the Carver Homestead Monument, about one mile south of Beeler, Kansas. The site of Carver's homestead is marked by a simple stone marker holding a plaque.

Finally, the George Washington Carver Museum is on the grounds of the Tuskegee Institute National Historic Site located at 1212 West Montgomery Road in Tuskegee, Alabama. The museum has exhibits and films and is open from 9:00 A.M. until 4:30 P.M.

## Irving Langmuir

Irving Langmuir, the first industrial chemist to win the Nobel Prize, proved to corporations and the government that investing large sums of money in unrestricted basic research could pay great dividends. In addition to publishing more than two hundred scientific papers, Langmuir is credited with sixty-three patents, including the mercury-condensation vacuum pump, the nitrogen-argon filled incandescent light bulb, atomic hydrogen welding, and an entire family of high-vacuum radio tubes.

Irvin Langmuir was born into a prosperous Brooklyn, New York, family in 1881, the third of four children of Charles Langmuir and Sadie née Comings. His parents always encouraged Irvin to carefully observe the natural world around him and to keep a detailed record of what he saw. Unfortunately, Irvin's ability to observe was hampered by poor eyesight, and not until age eleven, when the problem was detected and corrected, did he enjoy a clear view of his surroundings. Irvin's older brother Arthur, a research chemist, encouraged his curiosity, patiently answered his many questions, and helped him set up a laboratory in the corner of his bedroom.

Irvin attended elementary school in Brooklyn, then moved with his family to Paris for three years where his curiosity was stifled by rigid instruction. Back in the United States, he entered the private Chestnut Hill Academy in Philadelphia, where the flame of his intellect was rekindled. After attending the Pratt Institute's Manual Training High School in Brooklyn, he earned a degree in metallurgical engineering from the Columbia University School of Mines. As was the custom for physical scientists of the era, Langmuir went to Europe for his graduate work and wound up at Göttingen University in Germany where he worked under Nobel laureate Walther Nernst. His research focus was the behavior of gases produced in the presence of a hot platinum wire, a topic that laid the foundation for many of his later discoveries. After earning his doctorate in 1906, Langmuir taught chemistry at the Stevens Institute of Technology in Hoboken, New Jersey. One summer, he had the opportunity to work at General Electric's laboratory in Schenectady, New York. The lab's director, recognizing Langmuir's potential, offered him a job where he could devote all his time and energy to pure research and promised him adequate funding, a staff to assist in the work, and the freedom to pursue his own interests. Langmuir accepted the position and stayed at G.E. for the rest of his entire career.

His first major contribution was improving light bulbs by discovering that the lifetime of a tungsten filament could be extended by filling the bulb with an inert gas such as argon and that the efficiency of the filament could be improved by winding it into a tight coil. Playing around with light bulbs led him to the discovery of atomic hydrogen, which he then used to develop the hydrogen welding process. His investigations of filaments in a vacuum and in various gases led to a study of thermionic emission, the flow of charged particles from hot metals. He was one of the first scientists to study ionized gases and coined the term "plasma" to describe the charged gas.

Beginning in 1919, Langmuir turned his interest to atomic theory. In his most noted scientific paper, "The Arrangement of Electrons in Atoms and Molecules," he set forth his "concentric theory of atomic structure" wherein all atoms seek to fill their outer electron shell. He proposed that the chemical activity of an atom depended on the number and location of the electrons, and he contributed to the understanding of valence, covalence, and shared electrons.

Langmuir's observations of thin films on tungsten and platinum filaments were followed by experiments with oil films on water. From this research, he formulated a general theory of how substances are absorbed on surfaces: the absorbing surface has a catalytic effect, which causes a chemical reaction in the film. He also established the existence of monolayers, surface films with the thickness of a single molecule or atom. For his work on surface chemistry, Langmuir was awarded the Nobel Prize in chemistry in 1932.

Langmuir contributed to the war effort during both world wars. During World War I, he helped develop devices to detect submarines. After the war, he applied the technology to improving the quality of sound recording, a project he collaborated on with the famous orchestral conductor Leopold Stowkowski. During World War II, he helped create protective smoke screens and methods for deicing airplane wings. This research led him to the idea of using dry ice pellets and silver iodide crystals to induce precipitation, a process called "cloud seeding." The effectiveness of this technique remains controversial today.

Langmuir was a tireless worker who had a low tolerance for making small talk, but he enjoyed vigorous outdoor activities like hiking, skiing, and flying airplanes. He even climbed the Matterhorn! He was productive well into his seventies and spent the last years of his life traveling the world with his wife. Irvin Langmuir died in 1957 at Woods Hole, Massachusetts, after a short illness.

### Visiting Information

The Irvin Langmuir house is located at 1176 Stratford Road in Schenectady, New York. The two-and-a-half-story brick house was built around 1900, and Irvin Langmuir lived here from approximately 1919 until his death. Both the interior and exterior of the house remain as it was during the Langmuir period. The house, listed on the National Register of Historic Places, is a private residence and not open to the public.

## Linus Pauling

Linus Pauling wrote more than five hundred scientific papers spanning the fields of chemistry, biology, and medicine. He wrote eleven books, won every major prize in his discipline, and founded the field of molecular biology. He is the only person to win two unshared Nobel Prizes—one for chemistry and the other for peace. *The New Scientist* magazine named him as one of the twenty greatest scientists of all time, and he is generally considered to be the greatest chemist of the twentieth century.

Pauling was born on February 28, 1901, in Portland, Oregon, to Herman Pauling, a self-taught druggist of German descent, and Isabelle (Belle) Pauling, a descendant of an Irish pioneer family. Herman recognized his son's intellectual gifts and wrote a letter to the *Oregonian* newspaper seeking suggestions on a book list for Linus, who was a voracious reader. Soon after, when Linus was only nine, his father died suddenly, leaving Belle to support Linus and his sisters. Within a few months, Belle bought a house and supported the family by running it as a boarding house. When Linus was fourteen, a friend showed him a toy chemistry set. Mesmerized by the odorous fumes and colorful flames, Linus ran home and, finding equipment from an abandoned steel plant, began piecing together a primitive laboratory in the corner of his basement. Linus spent many happy hours during his teenage years doing experiments in his little laboratory.

By age fifteen, Linus had earned enough credits to enter Oregon Agricultural College (now Oregon State University) but lacked two history courses required for a high school diploma. He asked the school principal if he could take the two classes concurrently during the spring semester, but his request was denied. He decided to enter college without a diploma. (Forty-five years later, after Pauling had won two Nobel Prizes, his high school, Washington High School in Portland, agreed to award him a diploma.) Pauling majored in chemical engineering and soon demonstrated that he knew more chemistry than some of his professors. During his senior year, the understaffed chemistry department asked him to teach a class called "Chemistry for Home Economics Majors," an experience that allowed him to hone his lecturing skills and meet Ave Helen Miller, whom he later married.

By the time he graduated in 1922, Pauling had decided to shift his focus from chemical engineering to chemical theory so that he could pursue one of the most important questions in chemistry: how did atoms bond together to form molecules? He chose to pursue his graduate studies at a new research school in Pasadena, the California Institute of Technology. There, he became familiar with an experimental technique called X-ray crystallography, a method that enabled chemists to determine the structure and configuration of atoms within molecules and crystals.

After earning his Ph.D. in 1925, Pauling spent fifteen months in Europe learning about the revolutionary new theory of quantum mechanics from the likes of Erwin Schödinger, Neils Bohr, and Werner Heisenberg. When he returned to Caltech as a faculty member in 1927, as one of only a handful of chemists in the world familiar with quantum theory, he applied the theory to the problem of chemical structure. During the 1930s, he began publishing a series of classic papers on chemical bonds; in 1939, he summarized his work in *The Nature of the Chemical Bond*, one the most important scientific books ever written.

During World War II, Pauling was offered the job of heading the chemistry division of the Manhattan Project. He declined the offer, not because he was against the atomic bomb but because he would have to move his wife and four children. Pauling worked for several divisions of the National Defense Research Commission and contributed to the war effort by developing explosives and rocket propellants, patenting an armor-piercing shell, and inventing an oxygen meter for submarines. After the war, influenced by his wife's pacifist beliefs, Pauling became a relentless crusader against nuclear weapons.

Pauling had earlier redirected his structural studies from inorganic molecules to large organic molecules, especially proteins, laying the foundation for the new field of molecular biology. In 1950, Pauling coauthored a paper proposing that long organic molecules tended to have a helical structure. The discovery of the double helix of DNA emerged from Pauling's insight. Some believe that Pauling himself would have made the discovery had it not been for meddling of the U.S. government. Pauling had been invited to speak at a scientific conference in London in 1952, but because of Pauling's left-wing politics, specifically his opposition to atmospheric nuclear testing, the U.S. State Department, acting on the advice of the House Un-American Activities Committee, revoked his passport. During his trip, Pauling had planned to visit Cambridge University to take a look at new, high quality x-ray diffraction photographs of DNA. Had Pauling been able to see the photographs, we may never have heard of two guys named James Watson and Francis Crick.

Pauling won the Nobel Prize for chemistry in 1954 for "his research into the nature of the chemical bond and its applications to the elucidation of the structure of complex substances." Afterward, his political activism increased. He courageously spoke out against cold war politics and in support of a nuclear test ban treaty. His opinions were very unpopular in certain segments of American society. The American Legion, for example, called him an "abettor of the communist line." In 1958, he and his wife presented a petition signed by eleven thousand international scientists to the United Nations to end nuclear weapons testing. In 1963, John F. Kennedy

Linus Pauling picketing the White House as part of a mass demonstration protesting the resumption of U.S. atmospheric nuclear tests.

and Nikita Khrushchev signed the Partial Test Ban Treaty, ending above-ground nuclear testing. On the day the treaty went into effect, Pauling was awarded the Nobel Prize for Peace with the citation: "Linus Carl Pauling, who ever since 1946 has campaigned ceaselessly, not only against nuclear weapons tests, not only against the spread of armaments, not only against their very use, but against all warfare as a means of solving international conflicts." Pauling's prize was met with derision in the press. The *New York Herald Tribune* called him a "placating peacenik" and *Life* magazine characterized the prize as "A Weird Insult from Norway." The Caltech chemistry department never formally congratulated him, although the biology department threw him a small party. This cold response from Caltech led to his painful resignation from what had been his academic home for more than forty years.

For the next decade, Pauling bounced around working at various universities and think-tanks. In 1966, Pauling received a letter from biochemist Irwin Stone that advocated the use of vitamin C to promote health. Convinced the idea had some validity, Pauling focused his research on the idea that optimal health could result from maintaining just the right amount of certain molecules in the body, a field he called "orthomolecular medicine." He considered vitamin C to be one of the most important of these molecules, took large daily doses of vitamin C himself, and encouraged others to follow his example. In 1986, he wrote *How to Live Longer and Feel Better*, a popular account of his ideas. His claims were met with skepticism from the medical community who regarded the idea that vitamin C could cure the common cold and even cancer as quackery. Undeterred by the criticism, Pauling founded the Institute of Orthomolecular Medicine in 1973 in Palo Alto, California, where he conducted research until his death from prostate cancer in 1994. He never succeeded in producing any credible scientific evidence supporting his obsession with vitamin C. In 1996, the institute moved to Oregon State University and was renamed the Linus Pauling Institute.

## Visiting Information

Linus Pauling's boyhood home is at 3945 Southeast Hawthorne Boulevard in Portland, Oregon. Pauling moved here when he was nine after the death of his father. This is the house his mother ran as a boarding house. Pauling is buried in the Oswego Pioneer Cemetery in Lake Oswego, Oregon.

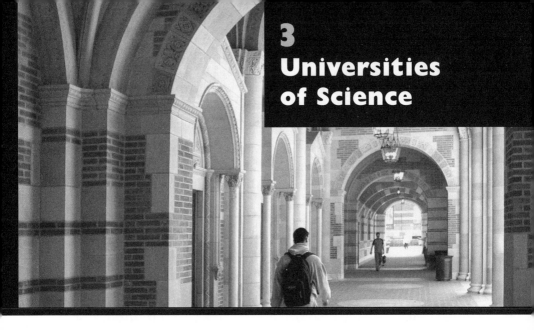

# 3
# Universities of Science

*Science discovery is an irrational act. It's an intuition which turns out to be reality at the end of it. I see no difference between a scientist developing a marvelous discovery and an artist making a painting.*

Carlo Rubbia

Now that we've met a few of the physicists and chemists who have made great contributions to our understanding of the universe, let's go visit the venues where they do their work. Most scientific research in the United States is done in one of three places: colleges and universities, government laboratories, and private institutes and industries. Our first stops are universities with especially strong reputations in the physical sciences. In chapter 4, we visit some big government-run laboratories. (Private laboratories aren't usually accessible to the general public so we'll have to skip those.)

University campuses are great places to visit, often elaborately landscaped and decorated with interesting art and architecture; they provide an idyllic setting for intellectual pursuits. Of course, an attractive campus is also a recruiting tool: if you're going to pay tens of thousands of dollars in tuition, it might as well be someplace beautiful. A stroll across campus brings back fond memories for those of us who have already donned our caps and gowns and takes us back to a time before families and financial responsibilities dominated our lives. On the other side of the ivy-covered walls, scientists are working with passionate intensity on all manner of research projects. Let's go behind the scenes and learn how science is done at a major research university.

When a newly minted Ph.D. scientist is hired by a university, it is on a probationary status, usually lasting seven years, during which time he or she must prove a worthy addition to the distinguished faculty. At the end of this time, each assistant professor must undergo a thorough evaluation process in which he or she is judged on several criteria, such as record of research, talent for teaching, and service to the university. At a small school, teaching is usually the most important factor, but at a big university research trumps teaching. The evaluation, done by a committee, usually takes several months. If the verdict is positive, then the professor receives a permanent job contract called tenure, the ultimate in job security. Denial of tenure is a very polite way of asking someone to leave. The idea is that tenure promotes academic freedom; tenured faculty can feel free to openly argue with campus administrators, hold and promote unpopular opinions, and pursue unfashionable topics. Young faculty members on the "tenure-track" are under pressure to be productive researchers.

Of course, research requires money for equipment, computer time, and salaries for research assistants. Where does the money come from? The main source of funding for scientific research in the United States is through various agencies of the federal government. The National Science Foundation (NSF), for example, has a budget of about six billion dollars, most of which funds basic scientific research. The Departments of Energy and Defense, along with NASA, are other important funding sources, particularly in physics and chemistry. But how does a researcher access this money? A scientist must write a grant proposal, an outline of the planned research. Money is limited so the competition for funding is fierce. The NSF receives more than forty-two thousand proposals annually, and less than 40 percent are successful. Depending on the source, a researcher's chance of getting funding is somewhere between 10 percent and 40 percent. The agency receiving a proposal subjects it to a "merit review." In this process, the proposal is carefully read by a committee of reviewers and ranked among other proposals in order of importance. The agency funds the projects in order of their ranking until the money is gone. If a proposal is rejected, then the researcher may revise the proposal and resubmit it or send it to a different funding source. Scientists spend a significant fraction of their time filling out all the required paperwork related to their grants.

After getting funding and actually doing the research, the final step is to communicate the results of the research to the scientific community. This is accomplished by writing a concise paper describing the research, including the technical procedural details, and submitting the paper to a scientific journal. There are literally thousands of journals dedicated to every imaginable scientific specialty and subspecialty. A physicist is likely to send a paper to a journal called *Physical Review*, a chemist to the *Journal of the American Chemical Society.*

When the journal receives the paper, it goes through a rigorous vetting process called "peer review." It works like this: The paper is read and critiqued by one or more experts, called "referees," who remain anonymous to the authors. The referees look for mistakes, problems with the procedure, and unsupported conclusions; then they recommend whether the paper should be published. Often, the recommendation comes with a list of corrections and clarifications. Peer review can be thought of as scientific quality control and is absolutely critical in guaranteeing the integrity of the scientific enterprise. If a paper is rejected by one journal, the authors usually just submit it to another journal. Sometimes, the authors plan out a pecking order, submitting the paper to the most prestigious journal first, then, if rejected, to the next most prestigious journal, and so on. Tenure committees take note of both the quality and the quantity of scientific papers a professor has published. A strong publication record goes a long way in swaying the committee to grant tenure.

Scientific walking tours through eight of America's most scientifically prestigious universities, listed in alphabetical order, are described below. Of course, nearly all major universities offer free student-guided tours of campus, but these tours are aimed mainly at prospective students and consist mainly of the guide walking around and pointing to different buildings. At the prestigious schools in this chapter, these tours are often crowded, especially during the summer. Also, the tours usually miss most of the scientific points of interest. (An exception is the MIT tour where, because the school is dominated by the sciences, the tour hits many scientific sites.) I chose these universities based on their scientific reputation and the number and significance of the physics- and chemistry-related sites that are accessible to the public. Before embarking on your stroll, be sure to have a campus map. By far the best campus maps I have found are the "Professor Pathfinder's" series. These detailed, color-coded maps include the campus and the surrounding area with the identity of each building printed directly on the building itself. So get out your map, enjoy your walk, and breathe in that rarified academic air.

## University of California at Berkeley

Berkeley is the land of hippies, radical politics, and an anything-goes atmosphere. The Free Speech Movement began here in the early 1960s, and the Berkeley campus hosted some of the first protests against the Vietnam War. The most infamous incident occurred on April 29, 1969, when 4,000 demonstrators clashed with 250 police officers over a block of land that had been unofficially designated as a "People's Park" by a group of left-wingers led by Jerry Rubin, Bobby Seale, and Tom Hayden. The police fired buckshot into the crowd, killing one rioter and blinding another. To quell the ensuing mayhem, Governor Ronald Reagan sent in the National Guard and for the next seventeen days, Berkeley turned into a war zone with repeated tear-gas-clouded clashes between students and the military. Most recently, a group of determined tree-sitters occupied the branches of a grove of oak trees in an effort to prevent bulldozers from clearing the trees to make way for an athletic training center.

Berkeley's roots go back to the 1849 gold rush when hundreds of thousands of prospectors poured into the state in search of quick riches. The drafters of the new state's constitution had a vision of establishing a university that "would contribute even more than California's gold to the glory and happiness of advancing generations." Twenty years later, the vision became a reality when the University of California opened in Oakland with ten faculty members and about forty students. The school's progressive stance began early: the next year, women were admitted. In 1873, the university, now with 200 students, moved a few miles north to a town called Berkeley, named for George Berkeley, an eighteenth-century Irish philosopher and bishop. Today, Berkeley is the flagship institution of the ten-campus, 220,000-student, University of California system.

Science has been strong at Berkeley since its inception when the distinguished naturalists Joseph LeConte and John LeConte were appointed to the original faculty. In 1888, the university thrust itself into the forefront of astronomy when Lick Observatory opened with the largest telescope in the world at that time. During World War II, three of the most important Manhattan Project scientists—J. Robert Oppenheimer, Ernest Lawrence, and Glenn Seaborg—worked at Berkeley. The myriad scientific discoveries at

Berkeley include the invention of the cyclotron, the creation of sixteen chemical elements heavier than uranium, and the identification of "good" and "bad" cholesterol and its role in heart disease. Berkeley's scientific excellence has resulted in fifteen Nobel Prizes in the sciences (eight in physics, seven in chemistry) by faculty members. The first went to E. O. Lawrence in 1939 for his cyclotron, and the most recent came in 2006 when cosmologist George Smoot won for his investigations of the early universe, which confirmed predictions made by the Big Bang Theory.

Begin your scientific tour of the Berkeley campus at the East Gate on Gayley Road. Walk down University Drive to the information kiosk where you can pick up a map. Just past the kiosk on your right is the Mining Circle. Follow the path opposite the circle leading between the buildings. The building on your left toward the end of the walk is Gilman Hall, former home of the chemistry department and a National Historic Chemical Landmark.

The story of Gilman Hall began when a highly regarded MIT chemistry professor named Gilbert Newton Lewis was lured to Berkeley to head-up the College of Chemistry. Lewis was promised a bigger budget, an expanded faculty and staff, and new facilities. Lewis began planning the "ultimate chemistry laboratory" that would be devoted to teaching and research in physical, inorganic, and nuclear chemistry. In 1917, Gilman Hall, named in honor of a former president of the university, opened its doors. The building is a two-story steel and concrete structure with a full basement mostly aboveground. A subbasement was later added to support Lewis's research in chemical thermodynamics. Lewis evidently liked the building because during the week, he sometimes lived here in a small bedroom and returned to his family in the country on weekends.

Research performed in Gilman Hall has resulted in two Nobel Prizes. First, in 1949 William F. Giauque won the prize for his work on how substances behave at extremely low temperatures. Then, in 1951, Glenn T. Seaborg was recognized for his discovery of elements beyond uranium. Four other chemists who did research here went on to win Nobel Prizes. Much of the research went on in the "attic." Enter the building, find your way to the attic, and locate room 307. This is where Glenn Seaborg and his colleagues, after using a cyclotron to bombard uranium with deuterons, identified plutonium as a new chemical element on a stormy Sunday night in February 1941. The next year, down the hall in room 303, Seaborg and two associates demonstrated that uranium-233 was fissionable. Both of these rooms, now used as offices, are marked with bronze plaques identifying them as national historic landmarks. During World War II, the Gilman Hall attic was fenced off for classified work in nuclear chemistry. As the chemistry department grew, the students and professors moved into new buildings. Today, Gilman Hall is occupied by the chemical engineering department, and the attic is still used for research.

Exit Gilman Hall and cross over to LeConte Hall, home of the physics department and several rooms of great historical significance. Ernest O. Lawrence built the first cyclotron, measuring only five inches in diameter, in room 329 in 1930. The room, now a classroom, is marked with a plaque. Arguably one of the most important meetings of the twentieth century took place in room 425 during June and July 1942. Here J. Robert Oppenheimer met with six physics "luminaries," as he jokingly called them, to brainstorm about how to make an atomic bomb and what the effects of an atomic explosion might be. The elite group included Hans Bethe, Felix Bloch, Emil Konopinski, Robert Serber, Edward Teller, and John Van Vleck. They outlined, on a microsecond time-scale, the evolution of an atomic explosion, and calculated the probable size, shape, and structure of a fission bomb. During the discussions, Teller brought up the idea of "The Super," a bomb based on nuclear fusion that would be vastly more powerful than a fission bomb. The fusion idea received serious attention, but ultimately the group decided that only the fission approach was feasible given the time constraint of a few years. Even then, building a fission bomb would require a Herculean effort. Teller's idea would have to wait. Teller also brought up the sobering possibility that the explosion might ignite the oceans or the atmosphere and incinerate the earth. After some calculations, the group became convinced that these concerns were unjustified. Part of the security measures installed for these top secret discussions was a wire mesh enclosing the balcony. The room is now a classroom, and the wire mesh, painted red, can be seen from outside the building.

Later that same year, Oppenheimer was named as the director of the Manhattan Project and performed the early planning and recruitment for the Los Alamos Laboratory from room 325. Oppenheimer's first office at Berkeley was room 219 ,and his attic office, room 426, connected to the luminary meeting room. Oppenheimer's home during his years at Berkeley was at #1 Eagle Hill in nearby Kensington, a short drive from campus. (See the entry on Oppenheimer in chapter 1.) On the street north of LeConte, you can find parking spaces marked "NL" for Nobel Laureate.

Return to your car and drive south on Gayley Road, turn onto Stadium Rim Way, and then onto Centennial Drive. Take Centennial Drive up the hill to the Lawrence Hall of Science (LHS). This is the public science center of Berkeley, named in honor of Nobel laureate E. O. Lawrence, the physicist who, as mentioned above, invented the cyclotron. The LHS, a leader in science education, is known for developing curricular materials used by science and math teachers across the country. Most exhibits here are geared toward kids, but some items will also interest adults. Most notable, a "Memorial Room" celebrates the life and science of E. O. Lawrence. Here, you can view a biographical film, examine a pair of "Dees" from one of his first cyclotrons, and admire his 1939 Nobel Prize medal. The seventy-five-ton magnet that Lawrence's thirty-seven-inch cyclotron was sandwiched between, sits in front of the building. The first weighable amount of plutonium oxide, isolated by Glenn Seaborg at the University of Chicago, is also on display. Other exhibits include the "Nanozone," which describes the emerging field of nanotechnology, the "Science on a Sphere," an interactive globe that displays scientific data from Earth, and an earthquake display with a seismograph connected to Berkeley's seismographic station.

Outside the building is *Sunstones II*, a fifteen-foot-tall granite astronomical sculpture with sight lines aligned with the solstices. The "Forces That Shape the Bay" explains the geological forces that affect the San Francisco Bay area, and you can't miss the giant sculpture of the DNA double helix. The LHS is situated on a hill and provides a breath-taking view of San Francisco Bay. Directly below the LHS is the Lawrence Berkeley Laboratory, described in chapter 4. The LHS is open daily from 10:00 A.M. until 5:00 P.M. Admission is $11 for adults, $9 for students and seniors, and $6 for children ages 3 through 6.

## Visitor Information

Of course, there is much more to see and do while visiting Berkeley. The scientific traveler may enjoy a leisurely stroll through the University of California botanical gardens about a half mile back down Centennial Drive from the LHS. The gardens are open daily from 9:00 A.M. until 5:00 P.M. There is a small admission charge. Free tours are available at 1:30 P.M. on Thursdays, Saturdays, and Sundays. On the main campus individual museums focus on anthropology, entomology, paleontology, and vertebrate zoology. The university offers free ninety-minute campus tours weekdays at 10:00 A.M. The tours leave from the Visitor Center, 101 University Hall. Weekend tours meet at the Campanile (Sather Tower) at the center of campus. The Saturday tour leaves at 10:00 A.M. while the Sunday tour departs at 1:00 P.M. If you prefer to explore at your own pace, a self-guided walking tour brochure is available at the Visitor Center or can be downloaded from the website below. Podcast tours and cell phone tours are also available. To get to the Berkeley campus from San Francisco, take the BART train to Berkeley Station and walk a block east on Center Street. From here, the Visitor Center is about two blocks north (left) on Oxford on the west (left) side of the street.

Websites: campus tours:
   http://visitors.berkeley.edu/free_tours.html
   Lawrence Hall of Science:
   www.lhs.berkeley.edu
Telephone: 510–642–5132

## California Institute of Technology (Caltech), Pasadena, California

Caltech is not known for its winning sports teams. The Caltech Beavers basketball team once suffered through a ten-year, 207-game NCAA Division III losing streak. But Caltech does much better in the arena of scientific discovery and accomplishment. Caltech faculty and alumni have won thirty-two Nobel Prizes, forty-eight National Medals of Science, ten National Medals of Technology, and five Crafoord Prizes. Famous Caltech professors include physicist Richard Feynman and chemist Linus Pauling. (For more on Feynman and Pauling, both Nobel laureates, see their entries in chapters 1 and 2.) Caltech is one of the most highly selective universities in the country with a student body consisting of only 900 undergraduate and 1,200 graduate students.

Caltech's history can be traced back to 1891 when Amos Throop, a Pasadena philanthropist, rented a building, hired six faculty, enrolled thirty-one students, and established Throop University. In 1907, astronomer George Ellery Hale, who had come to California to build the world's largest telescope on Mount Wilson, joined Throop's board of trustees. Following his personal credo "Dream no small dreams," Hale envisioned transforming Throop into a world-class "college of technology and science." Hale began by recruiting MIT chemist Arthur A. Noyes, enticing him with Pasadena's pleasant climate and the promise of a new chemistry building. Hale and Noyes realized they needed to attract an eminent physicist to the campus. They agreed that the University of Chicago's Robert Millikan would be a perfect fit and convinced the future Nobel laureate to join them in 1921. (For more on Millikan, see his entry in chapter 1.) Astronomer Hale, physicist Millikan, and chemist Noyes—a troika of genius known on campus as "Tinker, Thinker, and Stinker"—formed the first executive council of the newly renamed California Institute of Technology and set Caltech on course to becoming one of the world's leading research institutions.

Caltech has a delightful self-guided walking tour, "Along the Olive Walk," that describes past scientific discoveries and current areas of research inside its hallowed halls. The tour begins with a stop at the president's residence and then proceeds to the Atheneum. The first formal dinner held here was in honor of Albert Einstein who arrived at Caltech in 1931 for a three-month visit. The lounges and courtyards here are often used as movie sets. As you exit through the courtyard, you see a path lined with olive trees stretching out before you. The trees were planted to complement the Mediterranean feel of the architecture. In all, about 130 olive trees shade the campus walkways. Recently, a group of students and faculty decided to harvest the olives and sell Caltech olive oil in the bookstore. As you stroll through the trees, you see a cannon on your right. This relic of the Franco-Prussian War is fired to celebrate commencement and the last day of each semester.

The first science building on the tour is the Synchrotron Lab, which was home to an electron accelerator until it was dismantled in 1970. This is the building where the mirror for the 200-inch telescope at Palomar Observatory was ground and polished for ten years starting in 1936. Further along the walk, on your left, you come to a group of three buildings; Firestone, Guggenheim, and Karman make up the Aeronautical Laboratories. Inside the Guggenheim wind tunnels measure the aerodynamic properties of aircraft and automobiles. On the roof of the building is the T5 hypervelocity shock tunnel used for experiments simulating spacecraft entry into planetary atmospheres and aerodynamic braking. These laboratories helped establish a thriving aircraft industry in southern California and gave birth to the Jet Propulsion Laboratory. On the other side of the walk is the Thomas Laboratory of Engineering where engineers are working to make dams, buildings, and other structures more earthquake resistant.

Ahead a landscaped garden commemorates the site of Throop Hall, the very first building on campus. Throop had to be demolished after sustaining extensive earthquake damage in 1971. The rocks, grouped together by age and type, were selected by the geology faculty as representative of the vari-

Richard Thornton/Shutterstock

The Caltech cannon surrounded by olive trees.

ous kinds of rocks found in the nearby San Gabriel Mountains. A list of the rocks can be found on one of the large boulders at the front of the garden.

After climbing the steps a view of the nine-story Millikan Memorial building, home to the main campus library, greets you. The building on your left is really two buildings. On the left the Kellogg Radiation Laboratory houses a one-of-a-kind high-current, high-stability particle accelerator, custom designed by Caltech physicists to study nuclear astrophysics. Here William Fowler figured out how the chemical elements are cooked inside stars, a discovery that earned him the 1983 Nobel Prize in Physics. On the right side is the Sloan Laboratory of Mathematics and Physics. Here, Caltech faculty study nanostructures, ultratiny synthetic objects built from only a few hundred or a few thousand atoms. This is engineering on an atomic scale!

Along the walkway extending from the Sloan Lab is a bust of Robert Millikan who won the 1923 Nobel Prize in Physics for determining the electrical charge on an electron. Although he declined the title of "president," Millikan served as the administrative head of Caltech from 1921 through 1945. Students rub Millikan's nose for good luck before exams. On the opposite side of the central lawn behind a large tree stands a bust of the astronomer and telescope maker George Ellery Hale.

Back on the other side of the lawn to the side of the Millikan building is the Bridge Laboratory of Physics. Here all Caltech undergraduates take the required five semesters of physics, and presumably here Nobel laureate Richard Feynman delivered his famous lectures on physics. What was once the world's smallest motor can be seen in a display case in this building. To get there, enter the door marked "East Bridge" and go about halfway down the hallway. The display is on your left. This motor resulted from a challenge issued by Feynman, who offered a prize of $1,000 to the first person who could build a rotating electric motor that would fit into a cube measuring 1/64-inch on a side. The winner was Caltech alumni William McLellan who presented his motor to Feynman about two-and-a-half months after the announcement. Weighing in at a microscopic 250 millionths of a gram, the motor has thirteen parts that were machined using a watchmaker's lathe and pieced together under a microscope.

Continuing down on the same side of the lawn, you come to the Arms Laboratory of the Geological Sciences where current research examines the evolution of the Earth's climate and the motion of glaciers. The Arms building contains a cluster of clean rooms, affectionately known as "The Lunatic Asylum," wherein scientists examined the moon rocks returned by the Apollo astronauts. Next door, tucked behind a small courtyard is the Robinson Laboratory of Astrophysics. On the roof a dome once held a one-tenth-scale model of the 200-inch Hale reflector. The model tested the design of the telescope. Stepping back out onto the lawn, notice the decorative features on the surrounding buildings. These reliefs represent the scientific disciplines studied in the buildings at the time they were built.

The last building along this row is the Mudd Laboratory of the Geological Sciences. Geologists here study how ancient climatic conditions are revealed in trees, the natural magnetic compasses in birds and other organisms, and the behavior of the San Andreas fault. Around the corner a separate building, South Mudd, is home to Caltech's Seismological Laboratory. The study of earthquakes has a long and illustrious history at Caltech. The Richter Scale for measuring the magnitude of an earthquake was developed here in the 1930s by Beno Gutenberg and Charles Richter. Seismologists now use a different scale invented by the lab's former director Hiroo Kanamori. If you go inside South Mudd and climb up the stairs into the lobby, you find an exhibit on earthquakes and seismology along with a trio of seismographs.

Most remaining points of interest on the tour are in the areas of chemistry, biology, and computing. Of particular architectural interest is the archway between Church and Crellin laboratories. This archway is decorated with six figures: Pan with his pipes represents nature; a poet holding a tablet represents art; a figure carrying a human on his back represents energy; a figure lighting his torch from the sun represents science; a winged figure represents imagination; and a helmeted figure bearing tablets represents natural law. The reliefs were created by Alexander Stirling Calder, father of Alexander Calder, the famous mobile maker.

## Visiting Information

Caltech is located in Pasadena, California, about ten miles from downtown Los Angeles. The campus Visitors Center is housed on the first floor of the Office of Public Relations at 315 South Hill Avenue. Here you can pick up a campus map and a visitors brochure. Student-led tours of the campus are offered Monday through Friday and leave from the Office of Undergraduate Admissions at 355 South Hollistan Avenue. The self-guided walking tour described above can be downloaded from the Caltech website. The tour describes thirty points of interest on campus. Architectural tours are conducted on the fourth Thursday of September, October, and January through June. The November tour is on the third Thursday. These ninety-minute tours leave at 11 A.M. from the front hall of the Athenaeum at 551 South Hill Avenue. Be sure to pick up a copy of the Caltech newspaper *The California Tech* and enjoy the ultranerdy comics and the Dr. Quark advice column. The bookstore has a nice collection of science T-shirts. Visitors are required to display a parking permit if their car is parked on campus between the hours of 7:00 A.M. and 5:00 P.M. on weekdays. Permits are available at automated pay stations in the parking garages.

| | |
|---|---|
| Website: | www.caltech.edu (Click on "About Caltech" and find "Information for Visitors") |
| Telephone: | 626–395–6341 for information on student-led tours 626–395–6327 for information on architectural tours |

# University of Chicago, Chicago, Illinois

The University of Chicago boasts twenty-seven Nobel Prize winners in physics and fifteen in chemistry. In fact, more Nobel Prize winners (a total of eighty-two and counting) have been affiliated with the University of Chicago than with any other university in the world. (It is currently tied with England's Cambridge University; however, judging from recent trends, Chicago will soon eclipse Cambridge as the Nobel champion.) Here are just a few of Chicago's Nobel laureates and their discoveries: Albert Michelson, who measured the speed of light, in 1907 became the first American scientist to win a Nobel Prize; Robert Millikan determined the charge on an electron; Subrahmanyan Chandrasekhar did fundamental theoretical work on stellar evolution, neutron stars, and black holes; Enrico Fermi conducted experiments on artificial radioactivity; Willard Libby developed Carbon-14 dating; and Tetsuya Fujita devised the Fujita Tornado Scale (known as the F-scale) indicating the severity of a tornado.

The Chicagoans take their scholarship very seriously. For example, although the university was a charter member of the Big Ten Conference, the school's president eliminated the football program in 1939 citing the need to focus on academics rather than athletics. Football didn't return until 1969. Astronomer Carl Sagan, a Chicago graduate, tells the story of receiving a catalog from the university. Inside was a photograph of "fighting" football players. The caption under the photo read: "If you want a school with good football, don't come to the University of Chicago." Another photograph showed some drunken students. The caption read: "If you want a school with a good fraternity life, don't come to the University of Chicago." Sagan knew it was the right place for him. This extreme dedication to the academic life has lead some to jokingly refer to the University of Chicago as the school "where fun goes to die."

The University of Chicago was founded in 1890 by oil magnate John D. Rockefeller, who later called the university "the best investment I ever made." The land was donated by Chicago department store owner Marshall Field. William Rainey Harper, the university's first president, recruited a top-notch group of scholars to serve as faculty. As a later university president put it, "If the first faculty had met in a tent, this still would have been a great university." Harper created a new model for a university—a fusion of the American-style liberal arts college with the German-style research university.

In 1929, Robert Hutchins became the university's fifth president and introduced many curricular innovations for which the university is still famous. The Hutchins curriculum was an interdisciplinary approach where discussions replaced lectures, comprehensive exams replaced course grades, and classic books and original documents replaced textbooks. The core curriculum has evolved over time, but small group discussions and classic works remain standard features of a University of Chicago education. Few, if any, universities have had such a profound influence on the American higher education curriculum.

Begin your scientific walking tour of the University of Chicago campus at the massive stone gate directly across from the Regenstein Library on 57th Street. This gate is marked "Cobb Gate" on the campus map, but the arch has the words "Hall Biological Laboratories" carved into it. Walking through this gate is like walking back into time, a transition from the noisy modern world into a refuge for the intellect enclosed by ivy-covered edifices. The limestone buildings of the university's original quadrangles are adorned with the towers, spires, and gargoyles of the English Gothic architectural style, reminiscent of Oxford and Cambridge. Immediately on your left are the Botany Pond and the surrounding garden. In the summer with the plants in full bloom, this is the most beautiful spot on campus. Take a few minutes here to enjoy this tranquil setting. Just beyond the Botany Pond, turn left and walk through the column-lined arcade. The building on the right is Ryerson Hall, where Robert Millikan did his oil drop experiment and determined the electron's charge. The arcade leads into Hutchison Court, a lovely area designed by John Olmsted, son of the renowned landscape architect Frederick Law Olmstead. The court is bordered on one side by Hutchison Hall, a copy of Oxford's Christ Church Hall. Go inside and take a look at the dining hall modeled after those in England. In one corner of the court is Mitchell Tower, a scaled-down version of Oxford's Magdalen Tower. The main student center, the Reynolds Club, is also located here. Retrace your steps through the arcade and back out onto the main walk. Continue straight through the next gate and into the "Main Quadrangles." Take a leisurely stroll to the Harper Memorial Library at the far end of the quad. Enter the library, walk straight through, and exit onto 59th Street. Walk across the street to the Linné Monument and Readers Garden. Linné (more commonly known as Carl Linnaeus) invented the system of nomenclature that assigns a plant or animal to a particular class, order, genus, and species. For example, *Homo sapiens* is the genus and species name for human beings. Linné is portrayed here with some flowers clutched in his left hand. The Linné Monument sits in the vast green expanse of the Midway Plaisance. This is where the sideshow attractions were located during the World's Columbian Exposition in 1893. The use of the word "midway" to describe carnivals originates from this event.

Retrace your steps back through the Harper Library and into the main quad. Looking down the quad, make your way to the far left corner where you find yourself in front of the George Herbert Jones Laboratory. On August 18, 1942, in Room 405 of this building, a group of chemists led by Nobel laureate Glenn T. Seaborg, isolated 2.4 micrograms of plutonium, the first synthetic element. The effort was part of the Manhattan Project, and the plutonium was necessary for an atomic bomb. Go inside the building into the entryway where a display describes the sequence of steps leading up to the successful isolation of the tiny dab of plutonium. A series of twenty-five photographs can be viewed along with six instruments used to do the delicate job. If you are in an adventurous mood, the staircase leads to the fourth floor where Room 405 is tucked away off the main hallway. It is marked with a plaque designating it as a National Historic Landmark. The old wooden door, which is kept locked, probably appears as it did in 1942.

After exiting the Jones Lab, the walkway to your right leads between the buildings, out of the quad, and onto Ellis Avenue. The main campus bookstore is across the street to the left. Cross the street and turn right. After a short distance you see a sign for the Crerar Library. Holding more than 1.4 million volumes, the Crerar Library is one of the leading science libraries in the world. As you enter the library atrium, you see a suspended sculpture of aluminum and Waterford crystal. On display below the sculpture are the fossilized skull and armored plates of a supercroc. (You will have to ask the library attendant to allow you through the gate to view the fossils.)

Turn back to face Ellis Avenue, and then turn left past the Kersten Physics Teaching Center. Here are the main physics teaching laboratories. On top of the building is the dome of an observatory housing a small telescope used for instructional purposes. Continue down Ellis Avenue until you come to the Nuclear Energy Sculpture. This sculpture marks the site where a group of physicists, led by Enrico Fermi, built the world's first nuclear reactor. On December 2, 1942, the reactor was used to sustain a nuclear chain reaction. The reactor was built below the old Stagg Field, a site now occupied by the Regenstein Library. (The Nuclear Energy Sculpture and the history behind it are described in detail in chapter 7.) Across the street from the sculpture are the buildings housing the labs and offices for low-temperature physics, astronomy and astrophysics, high-energy physics, and accelerator physics.

Continue along Ellis Avenue until reaching 55th Street. Turn right and take 55th Street to Woodlawn Avenue. Turn right onto Woodlawn Avenue.

This street might well be considered "physicist row" because the former homes of three famous Nobel Prize–winning physicists can be found here. Each of the physicists and their homes are listed as a separate entry in chapter 1 of this book. Refer to those entries for biographical sketches. These privately owned houses are not open to the public. The first house you come to is that of Enrico and Laura Fermi at 5537 South Woodlawn Avenue; the Fermi family lived here from 1942 until Enrico's death in 1954. An informative "Chicago Tribute" marker is posted in front of the house. The large house at the corner of 56th and Woodlawn (5605 South Woodlawn, to be exact) is the Robert Millikan house, where he lived from 1907 until 1921; a National Historic Landmark Plaque can be seen on the entryway. Finally, Author Compton lived at 5637 South Woodlawn from the late 1920s until 1945; this house is not marked. Our scientific stroll through the University of Chicago and Hyde Park ends here.

Many other notable sights and museums in the Hyde Park area may be of interest to the scientific traveler. For example, if you continue down Woodlawn, you come to the Robie House, designed by Frank Lloyd Wright. Tours of the house ($12 for adults) are given several times a day. Earlier vintage Wright houses (not open to the public) dot Hyde Park; these include the Heller House at 5132 South Woodlawn, the Blossom House at 1332 East 49th Street, and the McArthur House at 4852 South Kenwood. Not surprisingly, the area is blessed with an abundance of bookstores. The Seminary Co-op Bookstore, a claustrophobic subterranean labyrinth holding more than 100,000 titles, has been called "the best bookstore west of Blackwell's in Oxford." Other sites include the Rockefeller Chapel, the Oriental Institute, and the Smart Museum of Art. A trip to the University of Chicago can be easily combined with a visit to the nearby Museum of Science and Industry described in chapter 9.

## Visiting Information

If you are in downtown Chicago without a car, the best way to get to the University of Chicago is by taking the Metra Electric Train (University Park Line). You can catch the train at the station across from Millennium Park and take it to the 55th–56th–57th Street stop or the 59th Street stop. Train schedules are available at the station. If you

Website: www.uchicago.edu
Telephone: 773–702–1234

have a car, there is plenty of street parking along the Midway Plaisance. There is no longer an official visitor's center, but campus maps can be found in Noyes Hall, the admissions office at 1101 East 58th Street, or the Reynolds Clubhouse. Maps are also available on the website. If you want to stay on campus, try booking a room at the International House.

## Columbia University, New York, New York

Columbia University, the fifth oldest institution of higher learning in the United States, was founded in 1754 as King's College under a royal charter of England's King Charles II. The school's purpose was to "enlarge the Mind, improve the Understanding, polish the whole Man, and qualify them to support the brightest Characters in all the elevated stations in life." With an enrollment of only eight students, the first classes were taught in a schoolhouse adjacent to Trinity Church in Manhattan. By 1767, King's College had established the first American medical school to grant the degree of M.D. The school closed for eight years during the Revolutionary War period but reopened in 1784 with a new name—Columbia College—since homage to the English king was no longer necessary. In 1849, the college moved from Park Place, near the current site of City Hall, to Forty-ninth Street and Madison Avenue. Here, the law school was established in 1858, and the country's first mining school, a precursor to the school of engineering and applied science, was founded in 1864. When Seth Low became president of the college in 1890, he consolidated the competing and independent schools under a central administration and made Columbia one of the nation's first centers for graduate education. In recognition of these changes, the trustees, in 1896, renamed the school by replacing "college" with "university." Low also coordinated a final move to the present Morningside Heights location in 1897. The campus here was modeled after the Athenian agora, a "place of assembly" where citizens would gather for military duty or to hear the proclamations of a king. In Athens, the agora featured a large rectangular open area bordered by grand public buildings.

The sciences at Columbia began to flourish in the early 1900s when Franz Boas transformed anthropology from a mere travelogue into a modern science and biologist Thomas Hunt Morgan introduced the chromosomal theory of heredity. In 1928, the Columbia-Presbyterian Medical Center became the first hospital in the country to combine teaching, research, and patient care. Physics at Columbia obtained international prominence in the 1940s with the atomic research of Enrico Fermi and I. I. Rabi. Seventy-six Nobel Prize winners have had an affiliation with Columbia University

The Low Memorial Library at Columbia University.

including thirty Nobel laureates in physics, nine of which were won for work done while a member of the department. These nine include Isidor Rabi, Charles Townes, and Leon Lederman. The walk outlined below takes you to Havemeyer Hall, home of the chemistry department, and Pupin Hall, home of the physics department. Both buildings have played important roles in the history of their respective disciplines.

Enter the heart of the campus along the "College Walk," which branches off of Broadway in line with 116th Street. The great dome of the Low Memorial Library, based architecturally on Athens's Parthenon and Rome's Pantheon, appears on the left. If the library looks familiar, you may have seen it in a movie (Hollywood directors have used it as a frequent backdrop). Actually, the building is no longer a library but houses administrative offices and the Visitors Center, located just inside the entrance should you want to join a campus tour. Behind Low Library and to the left is Havemeyer Hall, a red brick building trimmed with limestone that has been designated as a National Historic Chemical Landmark.

One of the six original buildings on what is now the Morningside Heights campus of the university, classes were first held in Havemeyer Hall, home base for the chemistry department, in 1898. Charles Frederick Chandler, a leading industrial chemist and the former President of the New York City health department was the driving force behind the construction of the facility. Funding came from sugar industry magnate and Columbia grad

Theodore Havemeyer, a good friend of Chandler's, who wanted to honor his father. Today, Havemeyer Hall is the heart of a three-building chemistry complex encompassing the Chandler Laboratories completed in 1927 and a six-story annex equipped with research and teaching laboratories.

Upon entering the building, the commemorative plaque telling of Havemeyer's landmark status is visible. Oddly, the entrance is on the third floor; research and teaching space for physical chemistry occupy the two lower floors. Find Room 309, the grand old lecture hall equipped with a domed ceiling and skylight, a 40-foot-long oak demonstration table, 330 tiered desks, and a gallery with a brass railing. The room, appearing much the same as it did in 1898, has also been used as a set for several movies, including *The Mirror Has Two Faces*, *Malcolm X*, and *Awakenings*.

Go up to the fourth floor where you find a long row of display cabinets lining one side of a hallway. This dusty, disorganized jumble of chemicals and antique scientific apparatus is a small fraction of the Chandler Chemical Museum collection; the remainder of the collection is in storage elsewhere in the building. The collection was started as a group of classroom instructional aides that Chandler used to liven up his lectures. His collection grew to include more than four thousand organic compounds (one thousand of which were first synthesized at Columbia) along with a wide variety of oils, explosives, pigments, and dyes. A set of early nineteenth-century apothecary jars is part of the collection as well as a group of items tracing the history of ceramics. Chemicals on display range from inorganic salts to organic chemicals used for medicinal purposes alongside other poisonous specimens. Hidden among the containers is one of the first samples of crude oil obtained from Texas oil fields. Some chemicals are labeled with cards describing their uses. More than half the display area is dedicated to old chemical equipment—everything from Bunsen burners, balances, and glassware to a distillation apparatus. Of particular interest is an apparatus for determining the alcohol content of beverages and another for testing sewer gas. It would be nice if someday this expansive collection could find a permanent home with enough space to display more items and a staff to organize and properly exhibit the objects as a fitting tribute to our chemical heritage.

Exit Havemeyer Hall, turn left, and then left again. Ahead is a high-rise brick building with an observatory dome on the roof. This is Pupin Hall, home of the physics and astronomy departments. The building is named in honor of Michael I. Pupin, a Serbian immigrant who graduated from Columbia and returned as an instructor of mathematical physics. He later became chairman of the department. In addition to being a scientist, Pupin was an inventor who developed a method of rapid x-ray photography and created a device called the "Pupin Coil" that increased the range of long-distance telephone calls. Under Pupin's guidance, this building was completed in 1925; after his death in 1935, it was renamed in his honor.

In the entry a bust of Pupin along with four plaques signifies the importance of this building in the history of twentieth-century physics. One plaque proclaims that the American Physical Society was founded at Columbia in 1899 and has been headquartered here since the 1920s. Another states that the Institute of Aeronautical Science was founded here in 1932. A third marker designates Pupin Hall as a National Historic Landmark stating vaguely that "this site possesses exceptional value in commemorating or illustrating the history of the U.S." So what happened here that earned national landmark status? In the basement of Pupin Hall Enrico Fermi, after learning of the splitting of a uranium nucleus by German scientists, successfully replicated the experiment on January 25, 1939. Thus, nuclear fission first took place on American soil, an event that lead to the Manhattan Project and the atomic bomb. After Fermi's success, work started on building an "exponential pile" consisting of a latticework of blocks made of graphite and uranium oxide. Members of the Columbia football team were recruited to move tons of the heavy material. The pile soon outgrew its space in Pupin and was moved to Schemerhorn Hall. After the Manhattan Project started in earnest in 1941, work on the pile shifted to Chicago and the research at Columbia refocused on the difficult problem of separating the isotopes of uranium. Much of the original Manhattan Project apparatus is still here, but sealed up in the basement and first floor of the building and inaccessible to the public.

A final plaque designates Pupin as a Historic Physics Site: "In recognition of Isidor Isaac Rabi for his discovery of the magnetic resonance method in 1938 and in recognition of the Columbia University Physics Department for its contribution to the advancement of physics." To explore the legacy of Nobel Laureate I. I. Rabi further, take the elevator to the eighth floor (the main entrance is on the fifth floor). Turn left out of the elevator and then left down the hall toward the physics library. On the left is Room 813 labeled with the name I. I. Rabi. Rabi's office, which he occupied for nearly fifty years, has been preserved as a memorial room in honor of the great physicist. To enter the room, go into the library and turn left. Here you'll find Rabi's desk and chair, blackboards, and many of his books. A set of five wall panels provides biographical information about Rabi, who played a key role in the development of radar during World War II and served as a science advisor to several U.S. presidents, the United Nations, and NATO. Nine of Rabi's students won Nobel Prizes.

In the hallway some interesting displays feature Columbia's Nobel Prize winners and other imminent physicists who worked here. Also on view is an assortment of mostly handwritten letters by such physics luminaries as Einstein, Millikan, Lorentz, and Planck. Particularly fascinating is a letter dating from 1938 written by Enrico Fermi in which he is expressing his desire to accept a position at Columbia, an invitation he had previously declined. Fermi, trying to escape the fascist powers in Italy, asks the department chairman to "invite me officially to teach at Columbia through the Italian embassy in the United States. Of course, you need not mention, or stress, that it would be a permanent position." Later in the letter, he implores: "Please do not give unnecessary publicity to this matter." After accepting his Nobel Prize in Stockholm, Fermi and his family secretly immigrated to the United States where Fermi had the Columbia job waiting for him.

## Harvard University, Cambridge, Massachusetts

Founded in 1636, only sixteen years after the pilgrims arrived at Plymouth, Harvard University is the oldest institution of higher learning in the western hemisphere and today ranks as the most prestigious university in the world. Established by a vote of the Great and General Court of the Massachusetts Bay Colony, the original purpose of the college was "to advance Learning and perpetuate it to Posterity; dreading to leave an illiterate Ministry to the Churches, when our present Ministers shall lie in the Dust." The site chosen for the school was a village called Newtowne, a carefully planned and stockaded village located three miles from the mouth of the Charles River so that it was safe from a naval attack. The legislators renamed the town Cambridge in honor of the English university where many Puritan leaders had been educated. The school is named after John Harvard, a minister who, in 1638, bequeathed half of his modest estate and all of his books to the college—a deal that surely ranks as the best bargain in the history of naming rights. In 1708, John Leverett became the first Harvard president who was not a minister, and he deftly steered the school away from Puritanism.

Harvard's rich scientific heritage can be traced back to John Winthrop, Harvard's first important faculty scientist. Winthrop, delivering lectures and performing demonstrations on electricity as early as 1746, was in charge of the first experimental physics lab in the country. Current and former Harvard faculty members have won forty-three Nobel Prizes, thirty-one of those in the sciences. Among the six Harvard Nobel Laureates in chemistry are Robert Burns Woodward for the laboratory synthesis of complex molecules, William Libscomb for the structure of boranes, and Walter Gilbert for developing methods for studying the structure of DNA. Nobel Laureates in physics—eleven in all—include Julian S. Schwinger for quantum electrodynamics, Sheldon L. Glashow and Steven Weinberg for unifying the electromagnetic force with the weak force, and Carlo Rubbia for experimental work in particle physics. Harvard has been especially prominent in the medical field, where faculty have won fourteen prizes. Nobel Laureates in medicine include James D. Watson for the structure of DNA, George Wald for the biochemistry of vision, and Joseph E. Murray for organ transplant procedures. Other significant medical breakthroughs include the introduction of surgical anesthesia, the development of the heart pacemaker, and the invention of the heart defibrillator. In 2001, physicist Lene Hau, building on her earlier work on slowing down the speed of light, stopped light completely by passing it through a cloud of supercooled atoms. To help you appreciate Harvard's scientific heritage, two self-guided walking tours are outlined below. The first takes you through the science area of the main campus, and the second takes you off-campus to some houses of great Harvard scientists.

Start your scientific walking tour of the main campus at the Johnstone Gate on Peabody Street. Proceed through the gate, and Massachusetts Hall is on your right. Completed in 1721, this is the second oldest academic building in the United States (the oldest is the Christopher Wren building at the college of William and Mary). During the Revolutionary War, the hall served as a barracks for the patriot army. Today, the ground floor holds the offices of the university president, while the upstairs holds freshman. On the other side of the walkway is Harvard Hall. Originally referred to as the "Philosophy Chamber," the building included a collection of scientific instruments known as the "Philosophical Apparatus," constituting what was probably the first experimental physics laboratory in America. John Winthrop, Harvard's first important scientist, was in charge of the apparatus, and the rooftop served as the college's observatory. Fire, sweeping through the original hall in 1764, destroyed the apparatus. Benjamin Franklin was asked to procure replacement equipment using his London connections with instrument makers. Some instruments Franklin purchased are on display in the Collection of Historic Scientific Instruments described below.

Ahead of you lies Harvard Yard, a pretty but plain expanse enclosed by stark, red brick buildings, shaded with trees but featuring no elaborate landscaping. Walk across the yard to pay homage to the statue of John Harvard. Turn left at the statue and head out of the yard and toward the entrance of the Science Center. The Science Center was built with a large donation from Edwin Land, a Harvard drop-out and founder of the Polaroid Corporation (some even claim that the building resembles an early Polaroid camera). Outside the entrance to the right is the Tanner Fountain that sprays a fine mist in summer and blows clouds of steam in the winter. Go inside where a sculpture titled *Topological III*, by physicist Robert Wilson, greets you. Walk down the hall to the central stairwell where you find on display the IBM Automatic Sequence Controlled Calculator (a.k.a. the Harvard Mark I), the first large-scale, automatic digital computer, arguably the machine that began the digital computing age. At the end of the hall to your left is the Harvard Collection of Historic Scientific Instruments. This wonderful collection of beautiful brass and wood instruments is a scientific "must-see" and merits its own entry in chapter 9.

After inspecting the instruments, exit the same entrance through which you entered, turn right, and walk around the building. The dark red brick building in front of you to your right is the Jefferson Laboratory, the first building designed specifically for physics research and instruction at an American university. In the mid 1800s, science was taught almost exclusively at special scientific and technical schools such as the Massachusetts Institute of Technology; science was not part of a liberal education that focused on the classics and theology. When Charles W. Elliott became president of Harvard in 1869, he proposed that the curriculum be expanded to include science and modern languages. The next year, John Trowbridge joined the physics faculty and argued that in addition to teaching, research must be a part of the mission of the physics department. In 1879, Trowbridge published a paper that revealed the results of a survey of laboratory apparatus across the United States; he discovered that the fledgling Johns Hopkins University had seven times the physical apparatus as Harvard. This finding helped Trowbridge convince Elliott that physics instruction should include a laboratory component and that expanding human knowledge through research should be part of the purpose of a university. To this aim, they decided that a new facility, designed for the dual purposes of research and laboratory-based instruction, should be constructed. Funding for the project came from Thomas Jefferson Coolidge, a Boston businessman who was a descendant of President Thomas Jefferson. It is in honor of Thomas Jefferson, an advocate for and contributor to science in America, that the building is named. The lab opened its doors in 1884, and within two decades, Trowbridge claimed that Harvard had the best equipment of any physics laboratory in the world. Early figures in the life of the lab include Edwin H. Hall who discovered an electromagnetic effect named in his honor, Theodore Lyman who made fundamental discoveries in spectroscopy, and Nobel laureate Percy Bridgeman. Perhaps the single most significant early achievement of the lab was establishing the science of architectural acoustics, a development brought about by President Elliott's request that physicist Wallace Sabine improve the sound quality in a lecture room in the newly built Fogg Art Museum.

Turn right, cross Oxford Street, and turn left. After a short walk you will find yourself standing in front of the Harvard Museum of Natural History where the biologist Stephen Jay Gould had his office. The natural history museum is a fusion of the Museum of Comparative Zoology, the Peabody Museum of Archeology and Ethnology, and the Mineralogical and Geological Museum. There is a $9 admission charge. The star attraction here is the Glass Flower collection, models of plant specimens made entirely from glass so that they would never decay and whither away. The models were created by a father and son team in Dresden, Germany, who worked continuously for nearly fifty years on the models. As you examine the specimens, you will find it difficult to believe that everything you see, including the delicate root systems, is made from glass. In addition to the plants, there are dozens of exquisite glass sea creatures including an octopus, a sea cucumber, and a jellyfish. The geological collection includes specimens of many chemical elements and a small display of meteorites.

As you exit the museum, look across the street and find the sharp angles of the Maxwell-Dworkin Computational Center. At the building's dedication, Steve Ballmer recounted the story of how on this site in the old Aikens Computation Center Harvard drop-out Bill Gates wrote the software code that became the first product offered by Microsoft.

Turn right and make your way around the museum building. The new facilities to your left are the high energy physics labs. You come out at the end of Divinity Avenue. Walk behind the Divinity School Dorm and into the courtyard of the Biological Laboratory, the home base of such Nobel laureates as James D. Watson, codiscoverer of DNA. This simple brick building has been transformed into a work of art by the architectural sculptures of Boston artist Katherine Lane Weems. The three pairs of green entry doors are decorated with eight gold sculptures representing the flora and fauna of the sea, air, and earth. The left doors symbolize the shellfish of the sea and include a starfish and a crab, the middle doors stand for the insects of the air and include a praying mantis and a bee, and the right doors show the plants of the earth and include the flower of a ginkgo tree. The entry doors are guarded by a pair of imposing rhinoceros. Encircling the top of the building are sculptures of animals representing the world's zoological regions. At the center is a symmetrical grouping of seventeen African elephants flanked by friezes of mostly individual animals including a gorilla, a lion, a polar bear, and a boa constrictor.

Make your way back out to Divinity Avenue and continue a short way down the street past the Peabody Museum. Turn right and then left into the courtyard of the chemistry complex. On your left bordering Divinity Avenue is the Fairchild Laboratory, a DNA research building. The balconies provide a safe refuge in case of an emergency. Across the courtyard is Mallinckrodt Lab, one of the first labs built in this area of the campus.

Make your way back out to Divinity Avenue and continue down the street until it ends at Kirkland Street. Turn left and the second street you come to on your left is Francis Street. Turn left and find 17 Francis Street. Now a National Historic Landmark, this was the home of Harvard geologist William Morris Davis, founder of the field known as geomorphology, the study of landforms. His most important contribution to the subject was the concept of a "cycle of erosion." This marks the end of the walking tour of the main campus.

If you are feeling energetic (this loop is about two and a half miles long), the second scientific walking tour takes you by several houses of Harvard scientists, all of which are National Historic Landmarks, but none of which are open to the public, and by the Harvard-Smithsonian Astrophysical Observatory. Start this tour at the information kiosk in Harvard Square. Facing the Kiosk, find Brattle Street which veers off to the right. After a block, Brattle Street takes an abrupt right. A scenic stroll up Brattle Street will take you by some beautiful old homes, including the Henry Wadsworth Longfellow National Historic Site, which you can tour.

Before reaching the Longfellow House, you will see Hawthorn Street on your left. Walk down the street and find 23 Hawthorn Street. This is the former home of Reginald A. Daly, an eminent Harvard geologist. Realizing the importance of field work in geology, Daly, beginning in 1893, carefully examined four hundred miles of terrain along the 49th parallel and formulated a theory explaining the origin of igneous rocks. An expedition to the Samoan Islands resulted in a theory relating sea level and coral atoll formation. Daly was the first to propose that the Moon was created when an enormous object impacted the Earth, a crazy idea at the time, but now widely accepted. He was also one of the first American geologists to support the theory of continental drift.

Return to Brattle Street, follow it to Craigie Street, and turn right. Look for Buckingham Street, the first street on your left. Walk up Buckingham Street until you see the Buckingham, Browne, and Nichols Elementary School on your right. Just before getting to the school, you find a side street, Buckingham Place, on your right. Walk down this street and find 10 Buckingham Place. From 1928 until his death in 1961, this two-and-one-half story frame house was the home of Harvard physicist and Nobel laureate Percy W. Bridgeman. The home now functions as a residence and faculty lounge for the elementary school. Bridgeman was a pioneer in the area of high-pressure physics. His Nobel Prize citation reads: "For the invention of apparatus for obtaining very high pressures, and for the discoveries which he made by means of this apparatus in the field of high pressure physics." This research eventually led to the creation of synthetic diamonds. Bridgeman also thought about and wrote extensively on the epistemology and methodology of physics, work that reached its pinnacle in his concept of "operational analysis," the idea that every physical concept is tied to the physical and mental operations by which it is measured and tested. After learning he had terminal cancer in 1961, Bridgeman committed suicide. A note in his pocket read: "It isn't decent for society to make a man do this thing himself. Probably this is the last day that I will be able to do it myself."

Go back to Craigie Street, turn left, and find 22 Craigie Street. This is the former home of Harvard mathematician George David Birkhoff, one of the preeminent American mathematicians of his time. In 1913, he proved Poincaré's "Last Geometric Theorem," a special case of the three-body problem, an achievement that brought him world fame. In 1931–1932, he developed his "ergodic theorem" that resolved a fundamental problem in the theory of gases and statistical mechanics and has also been applied to probability theory, group theory, and functional analysis. On the negative side, Einstein called Birkhoff "one of the world's great anti-Semites." During the 1930s, when many Jewish academics were fleeing Europe and trying to find jobs in America, Birkhoff allegedly tried to influence the hiring practices at American universities so that Jews would be excluded; some have defended Birkhoff by arguing that his anti-Semitic tendencies were not unusual during that period.

Continue down Craigie Street until it ends at Concord Avenue. Turn left on Concord Avenue and trudge up the hill. To your right appears the Harvard-Smithsonian Astrophysical Center for Astrophysics. Here, more than three hundred scientists from Harvard University and the Smithsonian Institution join forces to study the physical processes that govern the nature and evolution of the universe. On the third Thursday night of each month, the center's public "Observatory Night" features a nontechnical lecture and telescopic observing from the roof.

Turn right on Madison Street until you get to Garden Street. Turn left and look for 88 Garden Street. This was the home of Asa Gray, widely considered the leading American botanist of the 1800s. (The "Gray Gardens," named in his honor, are located across the street.) Gray was instrumental in classifying North American plant species and his illustrated *Gray's Manual* remains a standard reference in the field. He also founded the study of plant geography by noticing similarities between plant species in Japan and North America and suggesting that the species shared a common origin. On September 5, 1857, Charles Darwin wrote Gray a famous letter in which he outlined, for the first time, his theory of evolution by natural selection. Gray became an advocate for Darwin's theory and arranged for the first American edition of *On the Origin of Species*. Gray, a devout Presbyterian, argued that the theory of evolution was compatible with religious beliefs and attempted to reconcile the two in a collection of essays called *Darwiniana*.

Now go back down Garden Street for nearly a half mile toward Harvard Square until you reach Follen Street. (Look for the Longy School of Music building.) Walk down the street and find 15 Follen Street. This is the former home of Theodore W. Richards, a Harvard chemist and the first American to win the Nobel Prize in chemistry. He earned the award in 1914 "in recognition of his exact determinations of the atomic weights of a large number of the chemical elements." Richards was the first to show that a particular element could have different atomic weights, which supported the idea of isotopes. Other significant work included research into the compressibility of atoms, heats of solution and neutralization, and the electrochemistry of amalgams. His low temperature work on electrochemical potentials contributed to the development of the Nernst heat theorem and the third law of thermodynamics. Go back out to Garden Street to return to Harvard Square.

Another place near Harvard Square that might be of interest to the scientific traveler is the Mt. Auburn Cemetery, the country's first garden cemetery. Many famous people are buried here, including the following who made their mark in scientific or technical fields: geologist Louis Agassiz, mathematician Nathaniel Bowditch, telescope maker Alvan Clark, electrical engineer Harold "Doc" Edgerton, architect Buckminster Fuller, botanist Asa Gray, inventor Edwin Land, physicist and Nobel laureate Julian Schwinger, and psychologist B. F. Skinner. To get to the cemetery, walk up Mount Auburn Street (about 1.5 miles from Harvard Square) or take the #71 or #73 bus to Aberdeen Avenue. At the cemetery entrance, visitors can rent or purchase CDs or cassettes describing a one-hour driving tour and two seventy-five-minute walking tours; you can also rent an audio player.

### Visiting Information

Harvard University is located in Cambridge, Massachusetts, just across the Charles River from Boston. Driving and parking in Cambridge can be challenging so you might want to leave your car behind and take the Red Line subway to the Harvard stop. When you exit the subway station, look for the information kiosk near the out-of-town newspaper stand. Here, you can inquire about tour options. A good map of Harvard Square can be purchased for 25¢ or go to the Harvard Coop bookstore across the street and pick up a "Professor Pathfinder" map. Harvard offers twice-a-day (more often in the summer) free tours that give you a good orientation to campus. These tours leave from the Harvard Information Office just a block down Massachusetts Avenue.

## Massachusetts Institute of Technology (MIT), Cambridge, Massachusetts

If you asked somebody on a street corner to name a university known for science and technology, they would probably say "MIT." The Massachusetts Institute of Technology was the first American university to offer classes and curriculum in architecture (1868), electrical engineering (1882), sanitary engineering (1889), naval architecture and marine engineering (1895), aeronautical engineering (1914), meteorology (1928), and artificial intelligence (1960s). Seventy-one current and former members of the MIT community have won Nobel prizes. Among these are physicists Samuel Ting (1976), Wolfgang Ketterle (2001), and Frank Wilczek (2004), and chemist Richard Schrock (2005). Past scientific achievements include the development of modern food preservation techniques, the first chemical synthesis of penicillin and vitamin A, the invention of inertial guidance systems, high-speed photography, and the magnetic core memory used in computers. But MIT is perhaps best known for groundbreaking work in robotics and artificial intelligence.

MIT was the brainchild of William Barton Rogers, a distinguished natural scientist who recognized the need for a new kind of university that would prepare students for the challenges posed by the advent of the Industrial Revolution in America; challenges that a classical education did not address. Rogers developed a curriculum that stressed the practical and the pragmatic, where students would learn by doing through laboratory instruction. The state approved the founding charter for the university in 1861, but the first classes, held in a rented space in downtown Boston, were delayed until 1865, after the Civil War. The school's first building was completed the next year in Boston's Back Bay neighborhood and became known as "Boston Tech." In 1900, a proposed merger of MIT with Harvard was canceled after vehement protests from MIT alumni. In 1914, administrators announced a merger of MIT with Harvard's Department of Applied Science, but the plan was struck down by a state court. Meanwhile, MIT was outgrowing its Boston campus. When Richard McLaurin became the institute's president in 1909, he began searching for a new location. With money donated by George Eastman, MIT bought a mile-long stretch of industrial and swamp land along the Charles River in Cambridge. In 1916, MIT moved to its new location and began a new chapter in its history. During the 1930s, under the leadership of President Karl Compton (brother of Nobel laureate Arthur Compton) and Vice President Vannevar Bush, MIT rose to international prominence. Compton and Bush, revamping the curriculum, increased the emphasis placed on basic sciences like physics and chemistry and decreased the time students spent in shop and on drafting. The duo also attracted top-notch scientists and engineers, who contributed greatly to MIT's academic reputation. When Bush was appointed head of the Office of Scientific Research and Development during World War II, MIT became heavily involved in the war effort.

Today, MIT enrolls about four thousand undergraduates and six thousand graduate students. Although the sciences and engineering have tended to be male-dominated fields, about 45 percent of current undergraduates are female. About one thousand distinguished professors make up the faculty. MIT's arch rival is not nearby Harvard, but far away Caltech. When not studying, MIT students apply their genius-level IQs to devising elaborate practical jokes called "hacks."

Because MIT is dominated by science and technology, the free guided tours hit many scientific points of interest and provide a good way to see the university. The guided tour is described below along with a few sites you'll want to see after your tour. The tours begin in the rotunda of the Rodgers Building (a.k.a., Building 7—it may come as no surprise that all the buildings at MIT are numbered). While you wait for the tour to begin, check out the daily events calendar that is posted nearby. There might be something interesting to do. For example, a local theater often performs science-related plays. On my visit, I caught a performance of *QED*, a play about the life of MIT graduate Richard Feynman.

The tour begins with a glance at the inscription encircling the rotunda: "Established for Advancement and Development of Science and its Application to Industry the Arts and Commerce." Across the street is the Stratton Student Center, home base to the 200 or so campus clubs and organizations. Just about anybody can start a club for just about any activity; for example, the "Assassins Club" is a group of students who run around shooting each other with Nerf guns. A quick peek at a dorm room is followed by a stop outside the architecturally interesting Kresge Auditorium. Kresge's outer shell, an eighth of a sphere that floats free from the rest of the structure, is supported at only three points. Across the way is the MIT Chapel. Notice that the chapel, without windows, is partially lit through sunlight reflected into the interior by a moat. A hanging sculpture consisting of metallic pieces helps to spread the light throughout the chapel's interior. If there are no services, you may enter the chapel. If you're on the guided tour, you'll have to come back for a look inside.

Walking back across Massachusetts Avenue, the Harvard Bridge connecting MIT and Cambridge with Boston is off to the right. Time for the true story of Oliver Smoot, an MIT student who, in 1958, was used as a human yardstick. Smoot's buddies took it upon themselves to define a new unit of measure, called a "Smoot," as being equivalent to Smoot's height of 5 feet, 7 inches. They proceeded to measure the distance across the Harvard Bridge by having Smoot lay down, using paint to mark where his head came to, and then having him get up and move another Smoot. The distance across the bridge turned out to be 364.4 Smoots plus or minus an ear, the ear being a measure of the uncertainty in the measurement. The authorities have allowed the Smoot markings to remain because it provides the police with reference points to locate accidents on the bridge. In an ironic twist of fate, Oliver Smoot became chairman of the American National Standards Institute and president of the International Organization for Standardization.

Back you go into the Rodgers Building to peer down "The Infinite Corridor," a one-sixth-mile-long hallway linking several buildings. Twice each year, in May and October, the sun is aligned with the corridor so that sunlight shines down its entire length, forming what has been called "MIT-Henge." A short walk down the corridor and a right turn place you in front of a display case describing the 209 Robotics competition that was featured in the movie *21* (based on the true story of a group of MIT students who formed a blackjack team and won lots of money at Las Vegas casinos).

Get your camera ready because the next stop is Killian Court where you'll get a view of the famous Great Dome, the most familiar landmark at MIT. Notice the names of great scientists carved into the buildings to the left and right of the dome. Next is a meander through McDermott Court adorned by *The Big Sail*, a sculpture by Alexander Calder. Pass under the Green Building, the tallest building on campus designed by I. M. Pei. Traipse through the surreal Stata Center with its curved, cartoonish façade. The Stata Center occupies the site of MIT's legendary Building 20, a supposedly temporary timber-framed building constructed in 1943, but not torn down until 1998. Nicknamed the "magical incubator" by its occupants, it was a breeding ground for new ideas and innovative research and played a leading role in MIT's illustrious academic history. The first occupant of Building 20 was the MIT Radiation Laboratory, often referred to simply as the Rad Lab. Building on the discovery of radar by the British, the Rad Lab took over and further developed the technology, which became an extremely important weapon in World War II against the Axis powers. In fact, half of the radar deployed during the war was designed at the Rad Lab. Today, the Stata Center is home to pioneering research in computer science, linguistics, and philosophy. Next is a stroll down "Doc" Edgarton's Strobe Alley, a hallway with displays on Edgarton's use of high-speed photography. The tour finishes with a quiet glimpse into the Barker Engineering Library (notice the bust of Nicola Tesla hidden in a corner by the entrance) before returning to the Rodgers building.

MIT's surrealistic Stata Center, now occupying the site of the old Radiation Laboratory.

The guided tour does not take you to the Hart Nautical Collection or the MIT Museum. The Nautical Collection is just a short walk down the hall from the rotunda. Here, you find on display about thirty intricate ship models, including the *Mayflower* and a menacing Korean "Turtle" warship. Display cases explain the Department of Ocean Engineering and Naval Architecture research into structures, acoustics, and hydrodynamics. Of particular interest are the underwater robotic vehicles and robotic fish. Don't miss the "robo-tuna" used to study how fish swim. The Nautical Collection is free and open daily from 10:00 A.M. until 5:00 P.M.

The highlight of your MIT visit will no doubt be the MIT Museum, a short walk down Massachusetts Avenue. On your way, you pass a domed structure near some railroad tracks; this is MIT's nuclear reactor. The buildings behind it house the magnet lab and the Plasma Science and Fusion Center. The MIT Museum, located at 265 Massachusetts Avenue, merits its own entry and is described in detail in chapter 9. Don't miss it!

### Visiting Information

To get to MIT, take the Red Line subway to the Kendall Square/MIT stop or the Central Square stop. MIT offers free guided tours of the campus every weekday at 11:00 and 3:00. The Information Center, located just inside the Rogers Building at 77 Massachusetts Avenue, also has a self-guided walking tour that covers virtually the same route as the guided tour. Another option is to purchase the booklet *Art and Architecture at MIT: A Walking Tour of the Campus*, available at the List Visual Arts Center. The booklet provides you with detailed information about MIT's many sculptures and buildings by well-known artists and architects. Some of these sites are included on the guided or self-guided tours. The List Center is closed during the summer so you may want to order the booklet ahead of time.

## Princeton University, Princeton, New Jersey

On October 22, 1746, the Province of New Jersey, in the name of King George II, issued a charter establishing a college for "the Education of Youth in the Learned Languages and in the Liberal Arts and Sciences." The charter specified that "any Person of any religious Denomination whatsoever" could attend, a stipulation that was unique in the colonies at that time. Originally known as the College of New Jersey, the first classes, with an enrollment of ten students, were held in a parlor in Elizabeth, New Jersey. A decade later, the college moved into Nassau Hall in Princeton where, for the next fifty years, the entire college—dormitories, classrooms, and all—were housed. During the Revolutionary War, both British and American forces occupied Nassau Hall, which was hit by a cannonball during the Battle of Princeton. In 1783, the Continental Congress convened in the hall and thus it can be rightly claimed that the hall served, for a short time, as the U.S. capitol.

In 1896, during the celebration of its 150th anniversary, the College of New Jersey officially changed its name to Princeton University. A Princeton professor named Woodrow Wilson became president of the university in 1902 and began to grow the institution into a modern university. During his eight-year term, plans for the graduate school were finalized, the size of the faculty doubled, and an administrative structure was put into place. Wilson's curricular reforms included a general course of study for freshmen and sophomores coupled with more concentrated coursework for juniors and seniors. He also introduced small discussion classes called "preceptorials" that continue to supplement large lecture classes in the social sciences and humanities. After serving as president of Princeton, Woodrow Wilson served as the twenty-eighth president of the United States. Today, the Princeton campus is spread across 600 acres with more than 160 buildings. Over 1,200 full and part-time faculty teach nearly 5,000 undergraduates and about 2,300 graduate students, making it one of the smallest of the nation's leading research institutions.

Princeton's rich scientific tradition began when physicist Joseph Henry, codiscoverer of electromagnetic induction, arrived on campus in 1832. Henry's mere presence immediately elevated Princeton to the forefront of American physics. Henry's house and some of his scientific apparatus can be seen on the walking tour outlined below. Physicist Cyrus Fogg Brackett came to campus in 1873. A brilliant lecturer, Brackett wrote a popular physics textbook and advised the trustees on how to improve instruction in scientific subjects. Brackett also built up the university's collection of scientific equipment, a collection that had become outdated after Henry's departure in 1848. After Edison invented the light bulb, the mechanically talented Brackett rigged up an electrical system to make his the first classroom in America to be lighted by electricity. Physicist William Magie and mathematician Henry Fine, building on Brackett's foundation, formed Princeton into one of the world's leading centers for mathematics and theoretical physics.

Princeton faculty and alumni have garnered thirty Nobel Prizes, seventeen of those in physics. Among the Nobel laureates are faculty members Arno Penzias and Eugene Wigner and alumni Arthur Compton, John Bardeen, Richard Feynman, and Steven Weinberg. Mathematicians John Von Neumann and Kurt Gödel worked at Princeton as did John Nash, a Nobel laureate in economics whose struggles with mental illness were featured in the movie *A Beautiful Mind*. More recently, Andrew Wiles, a Princeton professor of mathematics, proved "Fermat's Last Theorem," one of the great unsolved problems in mathematics. But the most famous name associated with Princeton is that of Albert Einstein, who came to live and work here in 1933. Einstein was never a faculty member at Princeton University; instead, he worked at the Institute for Advanced Study, which is not formally associated with the university. Nevertheless, Einstein did enjoy a long friendship with the university and, for a while, had an office on campus. Many Einstein-related sites are part of the walking tours outlined below.

The scientific walking tour of Princeton is divided into an on-campus walk and an off-campus walk. Begin the on-campus portion of your Princeton tour at the Henry House, a large yellow Federal-style house on Nassau Street bordering the lawn in front of Nassau Hall. Take the path leading between the Henry House and the neighboring Caldwell House. Built in 1837, the house was designed by Joseph Henry, a professional physicist and an amateur architect. The house was originally located elsewhere on campus but has been moved several times to make room for new campus buildings. The north wing and south porch of the structure are more recent additions. Joseph Henry established Princeton as a leader in the field of physics, a distinction the university continues to maintain. Henry strung a telegraph wire between his office in Philosophical Hall and his home as a way of alerting the household that he would be coming home for lunch. The house now holds the offices for the Humanities Council. For more on Joseph Henry, see the entry in chapter 1.

Walk past the house, turn left and then right. On your left is the entrance to the Firestone Library, the world's largest collegiate library in which students have unrestricted access to the stacks. Most books are located underground to protect them from the sun's harmful ultraviolet rays. The library has changing exhibits that are open to the public so you may want to go inside to see the current displays. The exhibit area is to your right as you enter. Next to the library is the University Chapel, used mainly for weekly denominational and nondenominational services. Partly inspired by the chapel at King's College at Cambridge, this is the third largest college chapel in the world. Although founded by Presbyterians, the university has always been non-sectarian. Go inside to experience this beautiful space. In McCosh Court on the south side of the chapel is the Mather Sundial. Presented to the university in 1907, this copy of Oxford University's Turnball Sundial is one of the oldest pieces of free-standing sculpture on campus. For many years, tradition dictated that only seniors could sit at the base of the sculpture. Time has eroded that tradition, and now underclassmen can also claim spots.

Exit McCosh Court and walk between McCosh Hall and Murray Dodge Hall toward the Prospect House, a dining and social center for the faculty. Veer left, circling around the side of the Woolworth Music Building to the entrance of the Frist Campus Center. As the wording above the entrance indicates, this was originally the Palmer Physical Laboratory, dating back to 1908. Guarding the entrance are statues of Benjamin Franklin on the left and Joseph Henry on the right. Go inside, climb the stairs, and find Room 302. This room has been preserved to show an original physics lecture room. Notice the voltmeters above the board and the power supplies below. Exit the Frist Building, turn left, and find the entrance to Jones Hall (originally Fine Hall) which is attached to the campus center. Go in, turn left to the end of the hall and then turn right. At the end of the hall is Room 109. This was

Albert Einstein's office from 1933 to 1939 when the original offices of the Institute for Advanced Study were located on campus. Einstein also conducted work in the Palmer Physics Laboratory and often helped students with their math homework. The office is occupied by a Princeton professor so please respect professional privacy.

Leave Jones Hall, turn right and walk to Washington Road. Cross the street and turn right. The shiny metallic building at the corner of Washington Road and Ivy Lane is the Lewis Science Library. Go around the far side of the library, turn right and walk into the closed courtyard of the Jadwin physics building. Enter the building through the doors on your left. Just inside the entrance on your left is an electrostatic machine made under the direction of Benjamin Franklin and used by chemist Joseph Priestly. Around the corner to your left a set of display cabinets holds a fascinating array, including a glass rod with sealing wax used by Benjamin Franklin. Among the antique physics equipment, mostly electrical apparatus belonging to Joseph Henry, is an unexpected treasure: a voltaic pile brought to the United States by a Princeton professor in 1800, quite possibly the very first battery to appear in America. Artifacts related to Henry include his writing desk with a handwritten note, a model of the Earth's magnetic field built to demonstrate to students how the field could be produced by surface currents, an electromagnet and hammer used by Henry as a telegraph receiver, and an early electric motor. Some of Henry's electromagnets can be seen along with several of his induction coils. Look also for the Faraday voltmeter, an early electrical meter presumably designed by Faraday himself. This ends the on-campus part of the walking tour. Return to Nassau Street for the off-campus walk.

Begin your off-campus walking tour at the Bainbridge House located at 158 Nassau Street, home to the Historical Society of Princeton. In 2003, the Institute for Advanced Study donated sixty-five pieces of Albert Einstein's furniture from his Mercer Street home to the Historical Society. A few of these items are on display in the Princeton History Room. On my visit, Einstein's music stand could be seen along with a palm-sized puzzle game and a compass. Could this be the compass that was given to a young Einstein by his uncle, the compass that piqued Einstein's scientific curiosity? The museum attendant was unsure. The Historical Society has recently purchased an additional house, and, with this additional space, it may be possible to display more of the furniture collection in the future. (I certainly hope so! This collection should be out where the public can view it, not held in storage.) The Bainbridge House hours are from noon to 4:00 P.M. Tuesday through Sunday. The society offers various historical tours of Princeton on Sunday afternoons.

Just down the street at 102 Nassau Street is a clothing store called Landau. At the back of the store is a small "mini-museum" with Einstein-related photographs, documents, and historical information. When you exit the store, turn right and walk to the corner. Turn right on Witherspoon and walk to Wiggins Street (the public library is at the corner) and turn right. You see the Princeton Cemetery across the street. Go to Greenview Avenue and turn left. The cemetery entrance is at the end of the street. At the entrance, pick up a flyer detailing the history of the cemetery and offering a map with the locations of the graves of more than sixty notable persons who are buried here. Aaron Burr, pollster George Gallup, and President Grover Cleveland are among those interred here, along with several imminent scientists and mathematicians. Kurt Gödel formulated what is known as the "incompleteness theorem," a result that many consider to be the most important mathematical discovery of the twentieth century. Always eccentric, Gödel in his old age became obsessed with avoiding germs, compulsively cleaning his cutlery and wearing ski masks. At age seventy-two, Gödel died in a Princeton hospital from starvation because he refused to eat lest he ingest a few of the little bugs. John Von Neumann's famous contributions cross a wide range of mathematical fields, including game theory and computer science; he also made breakthroughs in quantum mechanics, worked on the Manhattan Project, and helped develop the hydrogen bomb. Von Neumann died at age fifty-three of cancer possibly resulting from his exposure to radiation. Gödel and Von Neumann are generally considered to be two of the greatest mathematicians of the twentieth century. Solomon Lefschetz was drawn to mathematics after losing both hands in an industrial accident. He made fundamental contributions in the areas of topology and differential equations. Renowned astronomer Lyman Spitzer pioneered the study of the interstellar medium and made major contributions to the area of plasma physics, including founding the Princeton Plasma Physics Laboratory. He was the first to suggest placing a large telescope in orbit and became a driving force behind the Hubble Space Telescope. NASA's Spitzer Space Telescope, which examines the sky in the infrared part of the spectrum, is named in his honor. Finally, Eugene Wigner won the 1963 Nobel Prize in physics "for his contributions to the theory of the atomic nucleus and the elementary particles, particularly through the discovery and application of fundamental symmetry principles." Wigner also worked on the Manhattan Project. Spitzer, von Neumann, and Gödel are buried near each other, and their graves are easy to find. Wigner's grave eluded me on my visit. Helen Dukas, Einstein's long-time secretary and housekeeper, is also buried here.

Make your way back to Nassau Street, turn right, and walk to the end of the street. Cross Bayard Lane (Highway 206) to the parklike area called Monument Drive. Here you find a bust of Einstein by Robert Berks. This is a copy of the bust from the Einstein Memorial in Washington, D.C. If you walk to the back of Einstein's head and look up you will see a pinpoint hole that allows light to penetrate the sculpture. According to Berk, "There is the light of the Big Bang."

Walk back down Nassau Street for about a block, and you see Mercer Street running diagonally. Walk down Mercer Street and find a two-story white house at 112 Mercer Street. This was Albert Einstein's home from 1935 until his death in 1955. See the entry in chapter 1 for more details.

The last stop on the off-campus walking tour is the Institute for Advanced Study, located about a mile from Einstein's house. You can walk to the Institute, as Einstein often did (Einstein never learned to drive), or take your car. To get to the Institute, continue down Mercer Street and take a left onto Olden Lane. The entrance to the Institute, Einstein Drive, appears on your right. (A sign says "Private Drive," but it's OK to visit as long as you don't go into the buildings.) Olden Farm at 97 Olden Lane serves as the home of the director of the Institute. Presumably, this is where Robert Oppenheimer lived during his tenure as director from 1947 through 1966. Continue around Einstein Drive to the front of Fuld Hall where you find visitor parking. The expansive lawn in front of the hall has a lane defined by rows of trees. There is a well-known photograph of Einstein walking alone down this lane. Any admirer of Einstein will want to retrace his steps.

The Institute for Advanced Study was established in 1930 by philanthropists Louis Bamberger and his sister Caroline Bamberger Fuld under the direction of Abraham Flexner. Original faculty members included Albert Einstein, Kurt Gödel, and John Von Neumann. The Institute is more commonly considered a "think tank." Here, unencumbered by teaching and other duties, scholars have complete freedom to pursue their own intellectual interests. The Institute has made major contributions to the fields of game theory, computer science, theoretical meteorology, string theory, and astrophysics, and twenty-two Nobel laureates have worked here. Today, 27 permanent faculty members work here alongside 190 visiting scholars.

Among the best known is physicist Edward Witten, a leading string theorist. Although the Institute is not formally linked to Princeton University, it enjoys collaborative ties to Princeton, Rutgers, and other nearby universities.

Another scientific destination you might want to pursue while in Princeton is the Princeton Plasma Physics Laboratory, a facility doing research on nuclear fusion. Unfortunately, there are no regularly scheduled public tours of this facility, but you can call ahead to possibly arrange a tour.

## Yale University, New Haven, Connecticut

Yale University, the third oldest institution of higher learning in the country, has a reputation for educating top national leaders including four signers of the Declaration of Independence, forty-five cabinet members, five hundred members of congress, and four of the last seven presidents. But Yale can also boast of leaders in the scientific fields including seventeen Nobel Prize winners in the sciences—three in chemistry, five in physics, and nine in medicine. In 1802, Benjamin Silliman taught the first modern science course (chemistry) in the United States, and in 1818 he founded the *American Journal of Science*, one of the oldest scientific journals in the world. In 1861, Yale awarded the country's first Ph.D. degree, and in 1876 Edward A. Bouchet became the first African American to earn a Ph.D. in America. Bouchet's degree was in physics, only the sixth doctorate ever awarded in that field.

Yale's charter was approved in 1701 for a school "wherein Youth may be instructed in the Arts and Sciences [and] through the blessing of Almighty God may be fitted for Publick employment both in Church and Civil State." One of the school's first benefactors was Elihu Yale, a Welsh merchant who donated a portrait of King George I, 417 books, and the money earned from the sale of nine bales of goods. In 1718, the institution that had been known simply as the "Collegiate School" was renamed Yale College in his honor.

Today, Yale, enrolling about five thousand undergraduates and six thousand graduate students from all fifty states and more than a hundred foreign countries, offers courses in forty scientific and engineering disciplines. Copying an arrangement used by Oxford and Cambridge, the undergraduate student body is divided into a dozen separate communities called "colleges" of about 450 students each. Each college has its own dormitory, dining room, and library surrounding a landscaped courtyard. A master, dean, and resident faculty members called "fellows," supervise the students within each college. This model, introduced in the 1930s, offers students the small, intimate atmosphere of a college within a large university.

The guided campus tours give a good overview of the history and culture of Yale and provide a peek at some pretty courtyards, inaccessible to the public. We begin our Yale tour with the guided tour and end it with a self-guided scientific walk that picks up where the guided tour leaves off.

The guided tours begin at the Yale Visitor Center. After a short talk about Yale's history, the student guide invites you to walk through the Phelps Gate into the inner sanctum of the "old college." Stop and pay homage at the statue of Nathan Hale, America's first spy. At the age of twenty-one, Hale was captured by the British and hanged. Before his execution, Hale uttered his most famous words: "I only regret that I have but one life to lose for my country." The statue stands next to Connecticut Hall, the oldest building in New Haven, completed in 1753 and designed after Harvard's Massachusetts Hall. Originally a student dorm, notable residents include Nathan Hale, Noah Webster, and Eli Whitney. The building is now home to the offices of the department of philosophy. Rub the foot of the Theodore Dwight Woolsey statue for good luck, as Yale students do before exams. Woolsey was the university's president for twenty-five years. After walking past Dwight Hall, the original library, gaze up at the gothic glory of Harkness Tower. The tower is 216 feet tall, one foot for each year of Yale's existence at the time it was built. Next, the guide unlocks the door leading into one or two of the cozy courtyards hidden in the interior of the student residence halls. The tour continues with a stop inside the Sterling Memorial Library. Created in the image of a Gothic cathedral, light shines through 3,300 hand-decorated windows, and the circulation desk resembles an altar, surrounded by academic, rather than religious, icons. Just outside the library is the *Women's Table*, a fountain sculpture by Maya Lin, who designed the Vietnam War Memorial when she was still a student at Yale. The numbers that spiral outward from the edge of the table give the number of female students at Yale until it equaled the number of male students (currently, the women outnumber the men). Journey across the street to the Beinecke Rare Book and Manuscript Library where you can gaze at a Gutenberg Bible printed around 1454. The library has an Exhibition Gallery, open to the public and often holding something of interest. The guided tour concludes in the rotunda of Memorial Hall. Be sure to take a peek at the Harry Potter-esque dining hall in the Commons adjoining the rotunda.

The gothic spire of Harkness Towers, a campus landmark at Yale University.

After thanking the student tour guide, you're ready to commence the scientific portion of the walking tour. This stroll takes you to the graves of several significant scientists, passes three scientists' former homes that have been designated as National Historic Landmarks, allows an optional stop at the Peabody Museum of Natural History, and concludes at the science area of campus. Ready? Exit the rotunda and turn left on Grove Street. Walk about a block to the entrance of the Grove Street Cemetery. Just inside is an office where you can ask the attendant for a cemetery map that locates the graves of more than eighty illustrious souls buried here. Among the departed are lexicographer Noah Webster, football legend Walter Camp, and band leader Glenn Miller. Inventors include Eli Whitney, who developed the cotton gin, and Charles Goodyear, who created vulcanized rubber. Notable scientists include thermodynamicist Josiah Willard Gibbs Jr., physics Nobel laureate Lars Onsager, mathematician and meteorologist Hubert Newton, science educator and chemist Benjamin Silliman and his son, Benjamin Silliman Jr., a geologist. (Note: The grave of physicist Josiah Williard Gibbs Jr. is easily confused with that of his father, Josiah Willard Gibbs Sr., because they are next to each other. The son's grave marker is long and coffin-shaped with the inscription "Professor of Mathematical Physics in Yale University, 1871—1903.") One-hour guided tours of the cemetery are given on Saturdays at 11:00.

After visiting the graves, exit the cemetery and turn left on Grove Street, retracing your steps. Turn left on Prospect Street, then right on Trumball Street until reaching Hillhouse Avenue. On the corner of Trumball and Hillhouse at 24 Hillhouse Avenue is the former home of geologist James Dwight Dana. Largely under Dana's leadership American geology grew from a science dominated by collecting and classifying into a mature science that attempted to explain the Earth's geological features.

Resume walking along Trumball Street until reaching Temple Street. On the corner of Trumball and Temple at 83 Trumball Street (currently a law office) is the former home of chemist Russell Henry Chittenden, generally considered the father of American biochemistry, who conducted pioneering research into the areas of digestion and nutrition. He lived in this house from 1887 until his death in 1943.

Go back down Trumball Street to Hillhouse Avenue and turn right. Hillhouse Avenue, a lane that Charles Dickens once called "the most beautiful street in America," is lined with impressive mansions that are now owned by the university. The president of Yale resides at 43 Hillhouse Avenue.

Hillhouse Avenue ends at Sachem Street. If you want to visit Yale's Peabody Museum of Natural History, turn right on Sachem. The museum is at the corner of Sachem and Whitney. The Peabody Museum has a Hall of Dinosaurs and a Hall of Minerals, Earth, and Space. Hours are 10:00 A.M. until 5:00 P.M. Monday through Saturday and from noon until 5:00 P.M. on Sundays. Admission is $7 for adults. Highlight tours are given on weekends at 12:30 P.M. and 1:30 P.M.

If you want to skip the museum for now, turn left on Sachem Street and walk to Prospect Street. In front of you, resembling the shell of a giant snail, is the Ingalls Skating Rink. Turn right on Prospect Street. To your right is the science area of the Yale Campus. When you get to the statue of Benjamin Silliman, turn right. The Sloan Physics Laboratory is on your right. Enter the building and turn left down the hall. Find the display case about physicist Josiah Willard Gibbs, whom some regard as the father of thermodynamics. The display holds a model of a Gibbs's governor and a plaster cast of the thermodynamic volume entropy surface for water. The cast was made for Gibbs by physicist James Clerk Maxwell, who was impressed by Gibbs's work. Exit the building to admire the façade of the Sterling Chemistry Laboratory, decorated with shields holding the chemical symbols of the elements. Return to Prospect Street and continue in your original direction. (Note: You may see a sign pointing to "Science Park." This isn't a "park" but rather an office building complex; there is nothing to see there.)

The house at 360 Prospect Street is the former home of paleontologist Othniel C. Marsh, who, along with his archrival Edward Cope, opened up the rich fossil deposits in the American west. Marsh was the first scientist to describe the dinosaurs known as stegosaurus and triceratops, and he was one of the first to support the idea that birds are descended from dinosaurs. Marsh's extensive collection of dinosaur fossils is housed in the Peabody Museum.

Turn around and go back down Prospect Street. On your left is the entrance to Leitner Observatory and Planetarium, used mainly for education and public outreach. Inside the building an exhibit presents several items of historical interest, including the first telescope in the western hemisphere used to observe Halley's Comet. Public viewing nights are, weather permitting, the first and third Tuesdays of the month. Next door, Farnam Memorial Gardens is a small park with a nice view of New Haven and the surrounding countryside—a good place to end your tour.

## Visiting Information

Yale University offers free campus tours at 10:30 A.M. and 2:00 P.M. on weekdays and at 1:30 on weekends. The seventy-five-minute tours leave from the Yale Visitor Center at 149 Elm Street. The Visitor Center has a room with a historical time line highlighting major events at Yale from 1701 to the present. Be sure to grab a campus map.

The musically inclined scientific traveler may also enjoy visiting the Yale Collection of Musical Instruments at 15 Hillhouse Avenue. The collection is closed during July and August, on Mondays, and during university holidays and recesses. It is open in the afternoons the rest of the year. Check the website (http://www.yale.edu/musicalinstruments/) for hours.

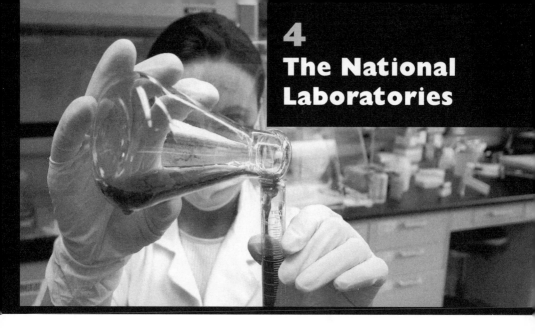

# 4
# The National Laboratories

*I have seen many phases of life; I have moved in imperial circles, I have been a Minister of State; but if I had to live my life again, I would always remain in my laboratory, for the greatest joy of my life has been to accomplish original scientific work, and, next to that, to lecture to a set of intelligent students.*

Jean Baptiste André Dumas

In addition to college campuses (as discussed in chapter 3), another major venue for scientific research, especially in physics, are the national laboratories. The current national laboratory system grew out of the Manhattan Project, the effort to build the atomic bomb. During World War II, secret labs were built at Los Alamos, New Mexico; Oak Ridge, Tennessee; and Hanford, Washington, so that scientists could conveniently collaborate to solve various problems related to the bomb. The labs were originally intended as temporary facilities, but, after the war, the newly created Atomic Energy Commission (AEC) took control and kept the labs alive. The laboratories at Los Alamos and Oak Ridge were designated as "National Laboratories" along with labs at Berkeley and Livermore, California; Brookhaven, New York; and the Argonne Forest near Chicago. These labs played a pivotal role in building nuclear power reactors, high-energy particle accelerators, and the hydrogen bomb.

Today, seventeen National Laboratories are funded through the Department of Energy (DOE). In fact, more than 40 percent of the total U.S. funding for the physical sciences comes from the DOE. Managed, operated, and staffed by academic institutions and private corporations under contract to the DOE, the labs employ more than 30,000 scientists and engineers. The labs perform research and development that because of the scope, cost, or multidisciplinary nature, cannot be done by a single university or corporation. Often, the labs are home to exorbitantly expensive equipment, machines, and supercomputers with price tags running into the hundreds of millions or even billions of dollars. Researchers from across the United States and around the world can apply to use the facilities. Labs with different missions normally cooperate with each other, but labs with similar missions often compete with each other for projects and funding. For example, Lawrence Livermore Lab was designed to compete with Los Alamos in the area of nuclear weapons research, the idea being that competition often results in higher quality work. The scientific scope of the labs has broadened from nuclear physics to include such fields as computers, meteorology, space, biology, alternative energy, and environmental science.

The four national laboratories described in this chapter (listed in alphabetical order) offer public tours. A few labs, not included here, may be toured in groups of ten. Fermi National Laboratory and the Stanford Linear Accelerator Center are described in chapter 5 on particle accelerators. Los Alamos and Oak Ridge National Laboratories are described in chapter 7 because they are best known for their role in the Manhattan Project. The National Renewable Energy Laboratory is described in chapter 8.

## Argonne National Laboratory, Chicago, Illinois

During the Manhattan Project, the University of Chicago's Metallurgical Laboratory (Met Lab), a band of about fifty scientists under the leadership of Enrico Fermi, built the world's first nuclear reactor on the campus of the University of Chicago (see the entry in chapter 7). About a year after the end of the war, for safety's sake, the Met Lab was moved to the Argonne forest about thirty miles from downtown Chicago, and the name was changed to Argonne National Laboratory, thus becoming the nation's first "national laboratory." Nuclear reactor pioneer Walter Zinn was appointed as the lab's first director and remained in that position for the next decade. In the early years, the lab was dedicated almost entirely to the development of nuclear reactors for military and commercial purposes. Here the reactors for nuclear-powered submarines were developed. In 1948, Zinn convinced the Atomic

can be used to determine whether a country, Iran or North Korea for example, intends to use its nuclear reactors to produce the material needed for nuclear weapons. Argonne is also home to two IBM Blue Gene/P supercomputers that make the lab one of the most advanced supercomputing centers in the world.

### Visiting Information

The Argonne Information Center is open to the public without prior arrangement. The center has informative panel displays on the lab's rich history, the ATLAS, the APS, and research projects. An electromagnet is on display along with a model of one of the accelerators. A giant model of a ridiculously complex molecule illustrates the kind of structural mapping of which the lab is capable. To get the most out of a visit, sign up in advance for a tour. Tours usually include stops at the APS, ATLAS, and the Engineering Research Exhibit, although sometimes visits to the APS and ATLAS must be omitted for various reasons. If you want to see everything, ask if all the sites will be included when you call to make tour arrangements.

Tours begin at the Information Center where you meet your guide for a drive around the lab. Various structures are pointed out along the route including the abandoned Chicago Pile #5 that was used as a neutron source in the lab's infancy. The first stop was the top floor of the APS building to view the circular structure that houses the x-ray beam. From this vantage point, several lab landmarks can be seen, most notably, the metallic dome of the historic Experimental Boiling Water Reactor (EBWR-1). Back on the ground floor, you are led past the control room and into a viewing gallery where the actual APS machine and several experimental stations can be seen. You might even see a scientist pedaling around the APS on a bicycle.

Surprisingly, the highlight of my tour was the stop at the Engineering Research Exhibit that showcases the lab's central role in the development of nuclear power. The exhibit holds reproductions of artifacts related to the first nuclear chain reaction, including lab notebooks and a bottle of Chianti used to celebrate the successful experiment. Be sure to touch a piece of graphite—not a reproduction—from the original reactor. The inner workings of a breeder reactor and a boiling water reactor are illustrated through large, interactive models. On display are many pieces of machinery and equipment related to nuclear power, including a fuel assembly. In all my scientific traveling, this is the best exhibit on nuclear energy I have seen! I have only one gentle criticism: Why not place this terrific exhibit in the Information Center so that it can be enjoyed at a more leisurely pace?

The Information Center is open from 9:00 A.M. until 5:00 P.M. Tours must be arranged in advanced by calling the number below. Public tours are normally held on a couple of Saturdays each month. Morning tours begin at 9:00 A.M. and afternoon tours begin at 1:00 P.M. Tours last about two and a half hours. Tours on other days of the week may be available; just call, and they will try to accommodate you. You must be at least sixteen years old to take the tour. A hotel, the Argonne Guest House, on the lab's grounds, welcomes visitors at reasonable rates. Argonne National Lab, located about thirty miles south of downtown Chicago, can be reached by taking the South Cass Avenue exit (exit 273A) off Interstate 55.

> Website:  www.anl.gov
> Telephone:  630–252–5562

## Brookhaven National Laboratory, Upton, Long Island, New York

In 1946, nine prestigious northeastern universities—Columbia, Cornell, Harvard, Johns Hopkins, MIT, Princeton, Penn, Rochester, and Yale—got together to establish a nuclear science laboratory that would house big, expensive machines that the individual institutions could not afford to build on their own. They chose as a site Camp Upton, a surplus army base on Long Island. The next year, the U.S. War Department transferred ownership of the site to the AEC, and thus Brookhaven National Laboratory was born. The lab has been home to three research nuclear reactors and a plethora of particle accelerators.

Through the years, Brookhaven's powerful mix of mind and machine power has produced five Nobel Prizes in physics and one in chemistry. The lab's first prize in 1957 resulted from the work done by T. D. Lee of Columbia University and C. N. Yang of Brookhaven. Using the Cosmotron particle accelerator, the physics duo showed that a principle of particle physics called parity conservation had been violated in some of their particle decay experiments. The 1976 prize was shared by Samuel Ting who used the Alternating Gradient Synchrotron (AGS) to discover a new particle called the J/psi particle which helped to confirm the existence of the charm quark. Four years later, James W. Cronin and Val L. Fitch, both of Princeton University, won for discovering a violation of a physics rule called CP symmetry; ironically, the pair had set out to prove that this rule is always obeyed.

The next Nobel came in 1988 for the discovery of a fundamental particle called the muon-neutrino. Columbia University scientists Leon Lederman, Melvin Schwartz, and Jack Steinberger used the AGS to produce a shower of particles called pi-mesons, which were then made to travel through a steel wall made of old battleship plates. The pi-mesons decayed into other particles as they made their way through the wall, but only the neutrinos made it to the other side of the wall and were detected. In 2002, Brookhaven's own Raymond Davis Jr. shared the prize for detecting solar neutrinos, particles produced in the nuclear fusion reactions in the core of the sun. Finally in 2003, Roderick MacKinnon, a visiting researcher from the Howard Hughes Medical Institute, won the Nobel Prize in chemistry for his explanation of how certain proteins help generate nerve impulses.

Today, Brookhaven employs about 2,600 people, operates on a budget of around half-a-billion dollars, and maintains a number of facilities that are used by nearly 4,000 scientists from across the United States and the world. The Relativistic Heavy Ion Collider (RHIC), a particle accelerator consisting of two crisscrossing rings 2.4 miles in circumference, uses more massive particles than Fermilab's protons or Stanford's electrons. Often, ions of gold are used. The RHIC is actually the final link in a chain of accelerators starting with a Tandem Van de Graaff. The ions move through the Booster and then are shot into the AGS before final injection into the RHIC, where the ions, traveling at 99.995 percent of the speed of light, make 80,000 orbits around the ring every second. Bunches of the ions are made to collide at six points around the rings where giant detectors are stationed to watch the collisions and measure the properties of the wreckage. The ions are so small that even though they are traveling very fast, the force of the ion collisions is about the same as the force of two colliding mosquitoes. The RHIC is designed to study matter as it existed only a fraction of a second after the Big Bang. According to current theory, matter at that time took the form of a quark-gluon plasma, a sort of perfect liquid.

Another major Brookhaven facility is the National Synchrotron Light Source (NSLS), which can be used to form images of everything from individual molecules to the atomic topography of surfaces. The very intense synchrotron light is created when electrons move in a curved path at nearly the speed of light. At the NSLS, this synchrotron radiation can be produced in the infrared, ultraviolet, and x-ray part of the spectrum. Scientists have used the NSLS to study nerve impulses, the crystal structure of nanomaterials, the chemical composition of bones to help understand arthritis and osteoporosis, and material dredged from New York harbor to measure pollution. Plans are in the works to upgrade the NSLS. The NSLS-II, with a beam 10,000 times brighter than the current NSLS, is scheduled for completion in 2015.

Other major Brookhaven facilities include the Center for Functional Nanomaterials (CFN) to study materials that may prove helpful in solving our energy problems. The NASA Space Radiation Laboratory studies the effects cosmic rays could have on spacecraft and astronauts, and the Accelerator Test Facility tests new concepts and components for particle accelerators. The Center for Translational Neuroimaging has instruments that can be used to explore the neurological causes of addiction, obesity, aging, and other conditions.

### Visiting Information

Brookhaven offers a series of "Summer Sundays" during July and August. Each Sunday program features a "Whiz Bang Science Show," a tour of one of the facilities, and other activities. Check the schedule on the website (click on community relations) to see which Sunday program appeals to your personal interests. Tours of the RHIC and the NSLS are usually available on one of the Sundays. No reservation is required; just show up between 10:00 A.M. and 3:00 P.M. Visitors age sixteen and older must present a photo ID.

> Website:  www.bnl.gov
> Telephone:  631–344–2651

## Lawrence Berkeley National Laboratory, Berkeley, California

Physicist Ernest Orlando Lawrence was sitting in the library one night when he happened upon a diagram in a journal article showing how high-energy particles might be created by giving them little electrical pushes with electrodes arranged along a straight line. Lawrence was captivated by the idea but realized that such a machine would be too long and cumbersome. He made the design more compact by sandwiching a circular chamber between the poles of a magnet. The magnet would exert a force on the particles causing them to move in a spiral rather than a straight line path. He called the device a "cyclotron" and finished piecing together a prototype in January 1931. The gadget—made of brass, wire, and sealing wax—was four inches in diameter and cost around $25. It worked and eventually earned Lawrence a Nobel Prize.

Lawrence was driven to build ever larger versions of his cyclotron. After the four-inch model came an eleven-inch, then a twenty-seven-inch. He needed a place to house the cyclotrons so his employer, the University of California, provided him with an old wooden civil engineering laboratory. Lawrence recruited a brilliant and dedicated band of about ten disciples, including his brother John, to run what became known as the Radiation Laboratory. Building bigger cyclotrons required money, and it was the depths of the Great Depression. Undeterred by the economic environment, Lawrence became a skilled fund-raiser and secured the financing he needed by touting the medical applications of his research. He kept his promises. With John Lawrence leading the way, the lab became the birthplace of nuclear medicine. A thirty-seven-inch cyclotron was operational in 1937 followed by a sixty-inch in 1939. When plans for a cyclotron greater than fifteen feet in diameter were announced, it became obvious that the machine was too big and potentially too dangerous to house on campus. The lab was moved to its present location in the Berkeley Hills above the university.

During World War II, the Radiation Lab played a key role in solving one of the most difficult problems associated with building an atomic bomb: how to separate the fissionable isotope of uranium, U-235, from the more common U-238. (See the entry on Oak Ridge for details.) Lawrence and his team converted the big cyclotron into a mass spectrometer that could separate the isotopes electromagnetically. Lawrence called the machine a "calutron" in recognition of the University of California. The calutrons were built at Oak Ridge, Tennessee, and eventually collected enough U-235 to make a uranium atomic bomb.

After the war, the AEC designated the Radiation Laboratory as a National Laboratory. When E. O. Lawrence died in 1959, the lab was renamed in his honor. During the 1950s, the lab's main focus shifted from classified weapons research to nonclassified basic scientific research. A group of scientists, most notably Edward Teller, split off from Lawrence Berkeley Laboratory and established the Lawrence Livermore National Laboratory whose mission was research on nuclear weapons.

Today, Lawrence Berkeley National Laboratory consists of seventy-six buildings spread out over two hundred acres, operates on an annual budget of nearly $600 million, and is managed by the University of California for the Department of Energy. Three thousand full-time employees work at the lab and another three thousand scientists from around the world visit the lab each year to participate in research projects or to use the facilities. Some fourteen hundred students, most from the Berkeley campus, receive scientific training at the lab. The lab has a multidisciplinary research program that encompasses nuclear physics, materials science, nanotechnology, computer science, earth science, environmental science, genetics, and climatology. Six user facilities are housed at the lab: the Advanced Light Source (a synchrotron radiation facility that generates intense light), the National Center for Electron Microscopy (a facility housing eight state-of-the-art microscopes that can image surfaces right down to the atomic level), the National Energy Research Scientific Computing Center, the Energy Sciences Network (a high-speed computing network), the Molecular Foundry (a facility used to study and build nanoscale materials), and the Joint Genome Institute.

The lab boasts a long list of major scientific achievements. A few recent accomplishments include the discovery of dark energy, construction of the world's smallest electrical switch made from a single Buckyball molecule, and the observation of ripples in the cosmic microwave background radiation that are thought to be the seeds from which the galaxies sprang. Sixteen chemical elements have been discovered at the lab, including all the transuranic elements from neptunium (element number 93) through seaborgium (element number 106). Two of these elements, californium and berkelium, are named after the lab's location and two others (lawrencium and seaborgium) are named after prominent scientists who worked at the lab. Eleven of the lab's scientists have won the Nobel Prize, including chemist Glenn T. Seaborg, physicist Luis Alvarez, and astronomer George Smoot.

## Visiting Information

The lab offers public tours, usually once every month on a Friday. Check the website for a schedule. You must sign up for a tour by calling the Community Relations Office at the number below. Tours begin at 10:00 and normally include a stop at the Advanced Light Source, which sits beneath the dome that held Lawrence's 184-inch cyclotron. Tour itineraries vary but may include visits to the National Center for Electron Microscopy, the Environmental Energy Technologies Division where scientists develop and evaluate methods of making buildings more energy efficient, the

> Website:  www.lbl.gov
> Telephone:  510–486–7292

Center for Beam Physics, which holds a laser-plasma acceleration lab, the Magnet and Cable Development Facility, the Center for Functional Imaging, which develops and improves imaging technology for the study of diseases, the Cancer Research Labs, the Genomic Sciences Lab, and the Energy Sciences Network Facility.

## Lawrence Livermore National Laboratory, Livermore, California

As the cold war heated up in the early 1950s, physicist Edward Teller, father of the hydrogen bomb, with the help of E. O. Lawrence, persuaded the AEC to create a second nuclear weapons laboratory. The new lab, initially a branch of the Berkeley Radiation Lab, would supplement the work at Los Alamos, provide the New Mexico lab with a little friendly competition, and subscribe to Berkeley's multidisciplinary team approach. The Radiation Lab's several large-scale projects, too big for Berkeley, had been built at a World War II Naval Air Base near Livermore, California, about thirty miles southeast of campus. In 1952, the Livermore Laboratory was established at the air base with thirty-two-year-old Herbert York, a former student of Lawrence, as the first director. Historically, the Berkeley and Livermore labs have enjoyed a close working relationship and were jointly administered until the early 1970s. Today, the Livermore lab has an annual budget of around $1.6 billion and employs more than 8,000 people, including 3,500 scientists, engineers, and technicians.

The lab's first major contribution was to shrink the size of nuclear warheads having an explosive energy equivalent to millions of tons (megatons) of TNT. Several of these smaller megaton warheads could be placed inside a single missile, which allowed it to hit several targets at once. The military term for this arrangement is Multiple Independently Targetable Reentry Vehicle (MIRV). In addition, the smaller warheads could be carried by smaller missiles that could be launched from a submarine. The submarine-launched Polaris missiles and their successors were key parts of the U.S. nuclear deterrent.

Aside from nuclear weapons, the lab conducts some of the world's most advanced scientific computer simulations. In 1953, the lab bought one of the first UNIVACs (Universal Automatic Computer) and has been at the very forefront of computer technology ever since. Whenever a new, improved supercomputer comes out, Livermore is often the first place to get one. Currently, the lab's IBM BlueGene/L, with 131,072 processors capable of performing over 280 trillion operations per second, is ranked as the most powerful computer in the world. The major use for this fantastic computing power is to simulate nuclear weapons performance. The United States no longer tests nuclear warheads directly; instead, scientists rely on computer simulations to make sure the nuclear arsenal is ready. In fact, Livermore, along with Los Alamos and Sandia Labs, is required to annually certify to the president that the nuclear stockpile is operational. This requirement drives the U.S. computer industry to develop increasingly advanced high-performance computers.

The most exciting and potentially world-changing new project at Lawrence Livermore involves nuclear fusion, the source of energy that makes the stars shine. In nuclear fusion, atomic nuclei are fused together and, in the process, energy is released. All nuclei have positive electrical charges, and, because like charges repel, getting the nuclei to fuse is not easily done. The trick is to get the nuclei very close together so that a short-ranged attractive force, called the strong nuclear force, kicks in and is able to overcome the electrical force of repulsion. Getting the nuclei close together requires very high speeds, which is to say, very high temperatures on the order of 100 million degrees Celsius. Such extraordinary high temperatures would melt any container holding the fusion fuel; also, if the fuel were to touch the container, then heat would flow from the fuel to the container, causing the fuel to cool down. Two methods of holding the fuel without a container (called "non-material confinement") have been devised: magnetic confinement and inertial confinement. Most nuclear reactors built to harness nuclear fusion, including those at the Princeton Plasma Physics Laboratory, have utilized the magnetic confinement system. Construction of the International Thermonuclear Experimental Reactor (ITER), a magnetic confinement reactor, is under way in France. Livermore is one of the few places in the world valiantly attempting to use the inertial confinement technique. The machine is called the National Ignition Facility (NIF), and at its heart is a spherical target chamber thirty feet wide. When a dime-sized deuterium (a form of hydrogen) cylinder is dropped into the target chamber, it gets zapped from every direction simultaneously by 192 powerful lasers. The deuterium absorbs 1.8 million joules of energy in a few billionths of a second. This is equivalent to about 500 trillion watts, almost 1,000 times the power generated in the entire United States over the same period of time. The laser blast compresses and heats the deuterium, thereby creating the densities and temperatures needed to ignite nuclear fusion. The NIF is the first inertial confinement system where the energy released by the fusion is greater than the energy used to initiate the fusion. Someday, the thermal energy unleashed in nuclear fusion could be used to boil water to make steam that will turn turbines and generate electricity.

What makes the NIF and nuclear fusion worth all the time, money, and resources? Nuclear fusion is the ultimate in electrical power generation for several reasons. First, because the amount of fusion is limited by the amount of fuel in the little cylinder, the reaction is totally controllable and completely safe; there is no chance of either an explosion or a meltdown. Second, no greenhouse gases are emitted, and no radioactive waste is created. (Some inner parts of the reactor become radioactive over time and need to be disposed of when the reactor is decommissioned, but this is a relatively minor problem.) The waste product from hydrogen fusion is helium. You could use the waste from a nuclear fusion power plant to blow up balloons and have a party! Finally, and most important, the fuel for fusion is hydrogen, the most abundant element in the universe. We have plenty of hydrogen here on Earth locked up in ocean water. The fuel is easily accessible and will last virtually forever.

In addition to fusion power production, the NIF has two other missions. First, because a thermonuclear warhead uses fusion to produce an explosion, the NIF can be used to study the physics of nuclear weapons, helping ensure that the stockpile is safe and reliable. Second, the NIF enables astrophysicists to study the nuclear processes that happen in a star. For example, the NIF can create the extreme conditions of pressure and temperature that occur in a supernova explosion.

### Visiting Information

For a top-secret weapons laboratory, Lawrence Livermore is surprisingly accommodating to visitors. The Discovery Center has displays describing the lab's activities in the areas of applied science, homeland security, and stockpile stewardship. The "History Tunnel" highlights people and achievements from the lab's illustrious past and a miniature version of the NIF can be viewed.

Livermore offers tours of its Main Site and of Site 300, an area used by the lab for testing nonnuclear explosives. Tour participants must be at least eighteen years old and wear flat-sole, closed-toe, and closed heeled shoes. U.S. citizens must register by calling the phone number below at least two weeks in advance. Foreign visitors may need approval from the Department of Energy.

Courtesy of Lawrence Livermore National Laboratory

The National Ignition Facility's spherical fusion target chamber, being gently lowered into place with one of the world's largest cranes.

The Main Site tours may include stops at the NIF, the Terascale Simulation Facility, home of one of the world's most powerful supercomputers, and the National Atmospheric Release Advisory Center (NARAC), which assists in emergency response decisions in case hazardous chemical, biological, or radiological materials are released into the atmosphere. Main Site tours are held on Tuesdays at 9:00 A.M., last about two and a half hours, and begin at the Discovery Center.

Site 300 tours may include stops at a western and northern vantage point for observing the site and its surroundings, the contained firing facility, and the environmental remediation facilities and wetlands. Site 300 tours are usually conducted on the first and third Friday of each month. The two-and-a-half-hour tours begin at 10:00 A.M. and leave from the Site 300 main parking lot off of Corral Hollow Road just east of Tracy, California.

> Website: www.llnl.gov
> Telephone for tour requests: 925–422–4599

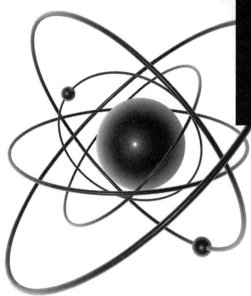

# 5
# Particle Accelerators

*As experimental techniques have grown from the top of a laboratory bench to the large accelerators of today, the basic components have changed vastly in scale but only little in basic function. More important, the motivation of those engaged in this type of experimentation has hardly changed at all.*

Wolfgang Panofsky

What is the universe made of? The first person to attempt to answer this fundamental question was the ancient Greek philosopher Democritus. He and his followers asked themselves: If I find some object, an apple or a blade of grass, and I take a very sharp knife and start cutting the object into smaller and smaller pieces, will I ever reach a piece so small that it can't be cut? In other words, is the number of possible cuts finite or infinite? Is matter smooth or is it grainy? Democritus argued that only a finite number of cuts were possible; at some point, you get to a piece of the object so small that it can't be divided. The Greek word for divisible is "tomos." Put an "a" in front of it, and you've got "a-tomos" or "not divisible." Atom.

We now know that atoms are in fact real objects, but we also know that what we call atoms aren't strictly atoms in the Greek sense of the word. We know that atoms can be subdivided into even smaller particles called electrons, protons, and neutrons, and we know that the protons and neutrons are made of still smaller particles called quarks.

The current state of knowledge regarding the fundamental particles that make up all matter is encapsulated in a theory unpretentiously dubbed the Standard Model. According to this well-established theory, everything is made of two kinds of particles, leptons and quarks. The quarks are always found trapped inside larger particles whereas the leptons can exist on their own. There are six kinds or, to use the physicists' terminology, "flavors" of quarks: up and down, strange and charm, top and bottom. Similarly, there are six kinds of leptons, the most famous of which is the electron. Each of these leptons and quarks has an antimatter twin—a particle with the same

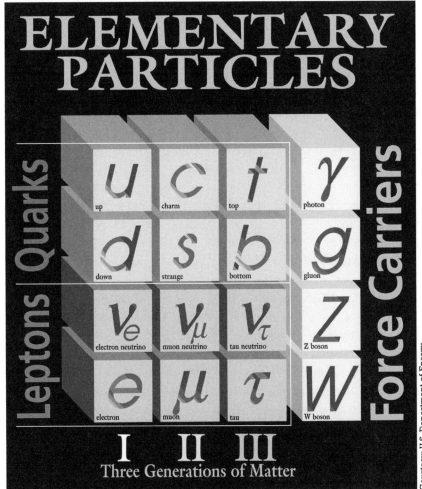

Chart listing and describing the twelve fundamental particles from which all matter in the universe is made.

mass, but opposite electrical and magnetic properties. For example, the antiparticle of an electron is called a positron. A positron has the same mass as an electron but with a positive charge.

Is the Standard Model the final word on fundamental particles? Maybe. But many physicists believe the current picture is too complicated and suspect that underlying all the particles of the Standard Model is something simpler. Currently in vogue in the physics community is the idea that all matter is ultimately made of objects called strings. These strings form closed loops like rubber bands. The different vibrational modes of the strings give rise to all the particles and forces we observe. Although string theory provides an elegant description of the universe, it has not yet been experimentally verified.

How do we know all this stuff about quarks and leptons? That's where particle accelerators come in. Just as astronomers use telescopes to observe outer space, particle physicists use accelerators to probe inner space. The first particle accelerator, called a cyclotron, was invented by physicist Ernest O. Lawrence in 1931. Before the advent of accelerators, physicists had to rely on cosmic rays, high-energy particles bombarding the earth from outer space, to study the structure of matter. With accelerators, physicists could produce their own artificial cosmic rays. Lawrence's first accelerator, only a few inches across, fit in the palm of his hand. By building larger accelerators, physicists can probe deeper into matter. The size of today's accelerators is measured not in inches, but in miles, making them some of the largest machines ever constructed.

What exactly is a particle accelerator (also known by its sexier name: atom smasher)? This machine accelerates beams of subatomic particles to speeds very close to the ultimate speed, the speed of light. The particles, accelerated by electric fields, can be steered into a circle by powerful electromagnets. There are two basic types of accelerators: linear accelerators and circular accelerators called synchrotrons. In a linear accelerator, a bundle of particles is shot in a straight line, receiving regular boosts of speed by strong electric fields. In a circular accelerator, the path of the particles is bent into a circle by using strong magnetic fields. The advantage of a circular accelerator is that the particle bundle can go around the circle thousands of times allowing its speed to approach the speed of light ever more closely. Higher speeds mean higher energies.

Two basic kinds of experiments are performed with particle accelerators: fixed target experiments and colliding beam experiments. In a fixed target experiment, a beam of particles smashes into a stationary target. In a colliding beam experiment, a bundle of charged particles is sent whizzing around in a clockwise circle, while a bundle of oppositely charged particles, moves counterclockwise. When the bundles have gained as much energy as possible, they are made to collide. Because higher energies can be produced with colliding beams than with fixed targets, it is the preferred experimental method on the frontier of particle physics.

The experiments produce two effects. First, the collisions bust open the particles so any smaller particles trapped inside get spewed out. The second effect is almost magical. Some of the energy of the collision is transformed into matter. How does that happen? $E = mc^2$. Recall that Einstein's famous equation says that matter and energy are different forms of the same thing. Thus, matter can be converted into energy (for example, a nuclear bomb), and energy can be converted into matter. By colliding energetic beams of particles, physicists can quite literally make matter. If you're not impressed by that, think of it this way: it's as if a head-on, high-speed collision between two ping-pong balls resulted in the creation of two bowling balls! The problem is that it takes a whole lot of energy to make just a little bit of matter. That's why accelerators are so big, the particles have to be going so fast, and the energies are so high.

A particle physics laboratory like those described below consists of not only the accelerator itself but also particle detectors and lots of computing capacity. To understand how the parts of the lab function together, consider the following analogy. Suppose you're trying to find a ball in a totally dark room. Luckily, you have a flashlight at your disposal. The flashlight produces a beam of light that you shine on the ball, the light reflects off the ball into your eye, and your eye sends the information to your brain for processing. In this analogy, the flashlight plays the role of the accelerator. The accelerator produces a beam of particles rather than a beam of light. The eye is the detector, collecting all the information and sending it to your brain, the computer, for processing.

Because particle physicists study the very smallest of objects, they need a conveniently small unit of energy. The unit of energy used in particle physics is called an electron-volt, abbreviated as eV. An electron-volt is defined to be the energy gained by an electron when it is accelerated through a potential difference of one volt. One thousand electron-volts is abbreviated as KeV, one million electron-volts is abbreviated as MeV, and one billion electron-volts is abbreviated as GeV.

The largest accelerator in the world is now the Large Hadron Collider (LHC), located at the Center for European Nuclear Research (CERN) near Geneva, Switzerland. It is regrettable that the United States missed its chance to continue to be the world leader in particle physics when—because of politics, cost overruns, and competition with NASA's International Space Station—the Superconducting Supercollider project was canceled in the early 1990s. In this chapter I describe the two main accelerator centers in the United States: Fermi National Accelerator Laboratory and the Stanford Linear Accelerator Center. Although these are two of the most famous scientific facilities in the world, the scientific traveler may be disappointed by visits. The problem is simple: the big accelerators and detectors are buried in underground tunnels to shield the apparatus from cosmic rays that would contaminate the beam. The protection works the other way, too. While the beam is on, radiation is produced. In fact, if you were to hug the pipes holding the beam, you would die almost instantly. The tunnels shield the outside environment from the dangerous radiation inside.

## Fermi National Accelerator Laboratory (Fermilab), Batavia, Illinois

Fermilab, named in honor of the great Italian physicist Enrico Fermi, is home to the second largest particle accelerator in the world, the Tevatron. With a circumference of four miles, the Tevatron can accelerate protons and antiprotons to 99.9999 percent of the speed of light and smash them together to produce collisions with energies of two trillion electron volts (that's where the "Tev" in Tevatron comes from). Three of the twelve fundamental particles in the Standard Model have been discovered here: the bottom quark in 1977, the top quark in 1995, and the tau neutrino in 2000.

The Tevatron is actually the final accelerator in a series of five separate accelerators that push the protons to within a hair's breadth of the speed of light. It all begins with an ordinary tank of hydrogen gas. Hydrogen, the simplest element, consists of a single electron bound to a single proton. The gas is introduced into a Cockcroft-Walton accelerator where an extra electron is added to the atoms to make negative ions. These ions are accelerated to an energy of 750 KeV, about thirty times the energy of the electrons in a television picture tube. The ions then enter a 500-foot-long linear accelerator called the Linac, which increases the energy up to 400 MeV. Before entering the next accelerator, the ions pass through a carbon foil that strips all the

Inside the tunnel of Fermilab's Main Injector, which accelerates protons and antiprotons before they are injected into the Tevatron.

electrons off and leaves only protons. The protons move into the Booster, a synchrotron-type accelerator 500 feet in diameter. They whiz around the Booster about 20,000 times while their energy is raised to 8 GeV. The next step is the Main Injector, which acts as a sort of proton switchyard by diverting some protons into a separate machine to make antiprotons (antiprotons have the same mass as protons, but a negative charge). With a circumference of two miles, the Main Injector accelerates both the protons and antiprotons up to 150 GeV and shoots them into the Tevatron. The protons go one way around the Tevatron, and the antiprotons go around in the opposite direction, thus circling the giant ring about 50,000 times each second. When the maximum speed is reached, the protons and antiprotons collide.

The particles smash together at the center of one of the two main detectors at Fermilab, the Collider Detector at Fermilab (CDF) and DZero. Each detector weighs about 5,000 tons, contains about 100,000 individual detection elements, and stands three stories high. Data is collected by the detectors and sent to the Feynman Computing Center for analysis.

The construction of Fermilab, originally known as the National Accelerator Laboratory, was approved by President Lyndon B. Johnson in 1967.

The small village of Weston, Illinois, was selected as the site of the new lab, and the village board agreed to vote Weston out of existence. The village houses have been converted into living quarters for visiting scientists or into small laboratories forming what is now "Fermilab Village." The lab's founding director, Robert Rathbun Wilson, committed the lab to the principles of scientific excellence, aesthetic beauty, environmental stewardship, fiscal responsibility, and equal opportunity. In addition to being a physicist, Wil-

Courtesy: Fermilab Visual Media Services

The Collider Detector, one of the giant particle detectors at Fermilab.

The main control room at Fermilab.

son was an artist and designed several sculptures that adorn the laboratory grounds. Today, Fermilab employs around 2,000 people and pays an annual electric bill of $12 to $18 million.

## Visiting Information

"Get to Know Fermilab" public tours are offered every Wednesday. No registration is required; just meet in the atrium of Wilson Hall at 10:30 A.M. These free ninety-minute tours take you up to the fifteenth floor where you can get a bird's-eye view of the lab and look at some displays, but you still don't get to see any machinery. To see some hardware, your best option is to register for an "Ask-a-Scientist" tour held on the first Sunday of each month. These three-hour tours, led by an actual Fermilab scientist, begin with a short talk and a visit to the fifteenth-floor viewing area, followed by a visit to the Linear Accelerator complex. Here, you see the main control room, the Cockcroft-Walton accelerator, and the linear accelerator gallery. What you see on the tour varies, but occasionally you visit a site related to the talk. These tours usually begin in the early afternoon. Check the website for a schedule. If you can't make one of the Sunday tours, just drop by and drive through the lab. A self-guided driving tour is described below.

Enter Fermilab at the Pine Street entrance off of Farnsworth Avenue. You drive under a blue-and-orange triple-span arch called "Broken Symmetry." The sculpture is asymmetric unless you stand directly under the point where the arches intersect and look up. From that vantage point, the sculpture looks perfectly symmetric. Continue down the road to the guardhouse where the attendant asks you for identification and the purpose of your visit. Be sure to ask for a map that indicates the laboratory's public areas. Immediately to your right is the Lederman Science Center named in honor of the Nobel Prize–winning physicist Leon Lederman, who succeeded Robert Wilson as Fermilab's director. Stop in here for a tutorial on high-energy particle physics. Although the center is directed toward children, the adult scientific traveler may enjoy a few exhibits. One exhibit invites you to bend a beam of charged particles with a magnet, thus demonstrating how the accelerator uses magnets to steer protons into a circle. One simple, but good, exhibit has pins of three different sizes. Push up on the pins to get an outline of your hand. The smaller pins give a better representation of your hand. This demonstrates that smaller probes bring out more detail. Also on display are a spark chamber, a scintillation detector, and a thirteen-inch bubble chamber showing the cryogenic piping and expansion tanks. A room marked "Ideas" has video displays on scale, powers of ten, fundamental particles and forces, and symmetry. Don't miss the exhibit on Leon Lederman just inside the entry. One of his journals is on display. The center has a sales counter where you can buy a Fermilab T-shirt and a few other items. Outside the center is a sundial. Stand on the month of the year, and your shadow hits on the time of day. The Science Center is open on weekdays from 8:30 A.M. to 4:30 P.M. and on Saturdays from 9:00 A.M. to 3:00 P.M.

Return to your car and turn right along the main road. After a short drive, Fermilab's dramatic, sixteen-story main office building, Wilson Hall, appears to your right. The design of Wilson Hall was inspired by the Beauvais Cathedral in France. In front of the building is a reflecting pool with a sculpture called *Hyperbolic Obelisk* at one end. The sculpture is made of three thirty-two-foot-tall stainless steel plates with the edges forming a hyperbola. Park your car and go inside Wilson Hall where you find yourself in one of the world's largest atriums decorated with a garden of trees and shrubs. At the center of the garden is a Foucault pendulum. A welcome desk, located in front of the garden, offers some literature about Fermilab. At the back of the building a cafeteria is open to the public. Go out the

Fermilab's Wilson Hall. *Mobius Strip*, sculpted by Fermilab's founding director Robert Wilson, in the foreground.

back door to admire a sculpture called the Möbius Strip mounted in the middle of a circular pool. As you look out from the building, the LINAC is beneath the ground to your right, and the Booster is directly in front of you. You are now standing on the roof of the Ramsey Auditorium. Admired for its superb acoustics, the auditorium serves as a venue for lectures and concerts.

Return to your car and head out along the reflecting pool. Directly ahead of you are some white power-line towers. Does the shape remind you of a famous number from math class? The towers are designed to resemble the Greek letter pi ($\pi$). Turn right at the intersection. To your right you notice a grassy berm running parallel to the road. The Tevatron lurks about thirty feet below this berm. On your left the semicircular Feynman Computing Center houses the main computing facilities for the laboratory. The precast concrete panels forming the south wall serve both a practical and an aesthetic purpose. They reflect sunlight to reduce solar heating, and they repeat the vertical lines of the Ramsey Auditorium. A bit further down the road on your left is the Technical Division building. Gracing the front of the building stands *Tractricious*, a sculpture made from scrap tubing left over from the construction of the Tevatron. Directly across the road, the CDF is hidden underground.

Continue down Road D to see the Buffalo Farm. Why does Fermilab have a herd of buffalo? (Actually, the correct name for these animals is American bison.) Robert Wilson started the herd in 1969 because he wanted to give local residents a chance to experience the natural environment of Illinois and give international visitors an opportunity to see a piece of Americana. If you want to get a peek at Fermilab Village, continue your drive along Batavia Road. If not, turn around and retrace your route. Before leaving Fermilab, you might want to take a leisurely stroll through

> Website:  www.fnal.gov
> Telephone:  630–840–5588
>             (for tour information)

the Prairie Interpretive Trail. The trail is part of a program to restore some of the lab's grounds to a native tall-grass prairie. Today, the Fermilab prairie is believed to be one of the largest contiguous tall-grass prairies in the world.

Fermilab is located in Batavia, Illinois, about forty-five miles west of downtown Chicago. Take the Farnsworth exit off of Interstate 88. The lab is open to the public from 8:00 A.M. to 8:00 P.M. daily from mid-April through mid-October and until 6:00 P.M. the rest of the year.

## Stanford Linear Accelerator Center (SLAC), Menlo Park, California

The two-mile-long accelerator operated at SLAC, the largest and most powerful linear accelerator in the world, produces bundles of electrons and positrons (particles with the same mass as electrons, but with a positive charge), accelerates them up to energies of 50 GeV, and smashes them together. The main accelerator, which lays claim to being the world's straightest object, is buried thirty feet under the ground and runs right under Interstate 280.

The accelerator has been operational since 1966, and today the lab employs about 1,500 people, including nearly 150 researchers with doctoral degrees. Physicists using the facilities at SLAC have been awarded three Nobel Prizes. In 1976, Burton Richter earned the prize for his discovery of the charm quark, one of the six quarks predicted by the Standard Model. Richard Taylor won the 1990 Nobel Prize for shooting electrons at protons and neutrons and proving that they were made of quarks. And in 1995, Martin Perl was honored for his discovery of the tau lepton, one of the six leptons included in the Standard Model.

SLAC is in a time of transition. Its primary focus is shifting away from high-energy physics toward photon science. As part of this change, the lab has constructed the world's first X-ray laser called the Linac Coherent Light Source (LCLS). This laser will produce ultrafast pulses of X-rays used to image single molecules. The LCLS can be used like a high-speed camera, which allows scientists to take stop-motion pictures of atoms and molecules in motion so that they can study the progress of chemical reactions. The accelerator mainly accelerates electrons, which then whiz around a storage ring and produce the X-rays.

### Visiting Information

At the time of this writing, SLAC had temporarily suspended its public tour program, although there is a Visitors' Center. According to the website, administrators plan to revamp the SLAC tour program so that it more accurately reflects the research done at SLAC. They intend to restart the tour program in 2009. SLAC is located in Menlo Park, California, south of San Francisco. Take the Sand Hill Road exit off of Interstate 280.

> Website: www.slac.stanford.edu

# 6
# Nuclear Weapons

*It is still an unending source of surprise for me to see how a few scribbles on a blackboard or on a sheet of paper could change the course of human affairs.*

Stanislaw Ulam

In 1939, two German scientists, Otto Hahn and Fritz Strassman, discovered nuclear fission, the process by which a heavy nucleus is hit by a neutron, causing the nucleus to split into two nuclei of intermediate mass. The two isotopes where fission is most easily accomplished are Uranium-235 and Plutonium-239. In a fission reaction, a tiny fraction of the original matter is converted into energy as prescribed by Einstein's famous equation, $E = mc^2$, where $E$ stands for energy, $m$ is mass, and $c$ is the speed of light. The equation means that matter and energy are different forms of the same thing; thus, matter can be converted into energy, and energy can be converted into matter. The speed of light is a very large number—300,000,000 meters per second to be exact. In the equation, the speed of light is squared, meaning that it's multiplied by itself. The value of $c^2$ is 90,000,000,000,000,000! So even if the mass, m, is very small, it gets multiplied by a humongous number, $c^2$, resulting in an output of enormous energy. That is, a little bit of mass can be converted into a whole lot of energy. The amount of energy released by the fissioning of a single U-235 is, on an atomic scale, huge; about seven million times the energy released in the explosion of one molecule of TNT. On a human scale, this really isn't a lot of energy. But what if you could get billions and billions of nuclei to fission? This is where the idea of a chain reaction comes in. Each fission reaction releases two or three neutrons. If more fissionable nuclei are close by, then the neutrons released in the first fission can, in turn, hit other nuclei causing them to fission, releasing more energy and more neutrons, which can then cause even more fissions, releasing even more energy, and so on. In a fraction of a second all this happens to release a prodigious amount of energy. This is the essence of an atomic bomb.

The discovery of fission by scientists in Nazi Germany did not go unnoticed in the United States. Leo Szilard, a Hungarian-born Jewish physicist who had immigrated to America, fully appreciated the potentially ominous implications of the German discovery. After all, Szilard had first come up with the idea of a chain reaction in 1933, while crossing a street in London. Szilard understood that nuclear fission might lead to the development of a new type of weapon with enough destructive energy to wipe out an entire city. He also understood the implications: if Hitler got the bomb first, then the Nazis would win the war and enslave Europe.

At the urging of his colleague Eugene Wigner, Szilard decided that he must warn the U.S. government about this possibility. In the summer of

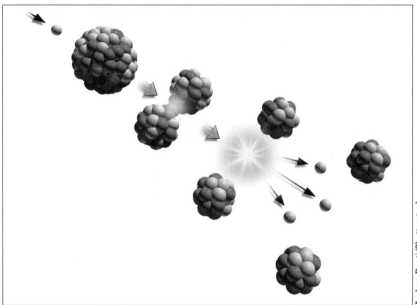

Andrea Danti/Shutterstock

Diagram illustrating a chain reaction.

1939, they arranged a meeting with the one scientist in the world who could command the attention of the president of the United States: Albert Einstein. They explained the problem to Einstein, and he agreed to sign a letter to President Franklin D. Roosevelt. The letter was hand delivered to the president by Alexander Sachs, a friend and confidante of Roosevelt. This letter, arguably the single most important letter written in the twentieth century, read in part, "This new phenomenon would also lead to the construction of bombs, and it is conceivable—though much less certain—that extremely powerful bombs of a new type may thus be constructed."

The United States finally got serious about beating Hitler to the bomb in December 1941, when war was declared on Germany and Japan. The project was placed under the control of the Army Corps of Engineers who had an office in Manhattan. Thus, the top secret, all-out, $2-billion effort to build the bomb became known as the Manhattan Project. General Leslie Groves was ordered to oversee the project. Groves, in turn, handpicked physicist J. Robert Oppenheimer to be the scientific director. The raw material for the bombs, uranium and plutonium, was produced at Oak Ridge, Tennessee, and Hanford, Washington. The bomb was designed and built at Los Alamos, New Mexico. Oppenheimer recruited to Los Alamos what is surely the greatest concentration of scientific genius the world has ever seen. The names read like a Who's Who of twentieth-century physics: Hans Bethe, Enrico Fermi, Edward Teller, and Richard Feynman are but a few. The scientists at Los Alamos worked on two bombs; one bomb used uranium, the other, plutonium. While the uranium bomb design was relatively simple and straightforward, the plutonium bomb proved to be a much greater challenge.

In July 1945, the plutonium bomb was successfully tested in the New Mexican desert about two hundred miles south of Los Alamos. The test was code-named "Trinity." By that time, Hitler and the Nazis, who served as the rationale for building the bomb in the first place, had been defeated. A talented group of German scientists, lead by Nobel Prize–winning physicist Werner Heisenberg, had indeed tried to build an atomic bomb. Luckily, as the Allies found out after capturing the scientists toward the end of the war, they didn't get as far along on the bomb as we feared they might. Several reasons explain their surprising lack of progress. First, the United States and the Allies did everything they could to deny Hitler the facilities and raw materials he needed to build a bomb. More significant, Hitler was impatient. He wanted new weapons that could quickly be used on the battlefield, and significant uncertainty surrounded when and even if an atomic bomb could be built. Hitler therefore concentrated his dwindling resources on Werner von Braun's V-2 rockets, rather than Heisenberg's atomic bomb. Nevertheless, we were still at war with the Japanese, who also had an atomic bomb project spearheaded by Japan's leading physicist, Yoshio Nishina. As World War II dragged on in the Pacific, the project lost momentum, and the Japanese team made less progress than the Germans.

Wanting to avoid a full-scale invasion of Japan at all costs, President Harry Truman decided to use the bomb against the Japanese. In anticipation of the use of the atomic bombs, several Japanese cities, including Hiroshima, Nagasaki, and Kokura had been removed from the air force bombing so that the atomic bombs could be used against a virgin target. This allowed a better determination of the bomb's destructive capabilities. In the early morning hours of August 6, 1945, a specially outfitted B-29 Superfortress took off from Tinian Island in the South Pacific. The pilot, Colonel Paul Tibbets, named the plane *Enola Gay*, after his mother. In the bomb-bay sat the untested uranium bomb, nicknamed "Little Boy." At 8:15:19 on a Sunday morning, the bomb was released. Exploding forty-three seconds later, nearly 2,000 feet above the target city of Hiroshima, the bomb released an amount of energy equal to 12,500 tons of dynamite. In an instant, the city of Hiroshima was almost totally destroyed. Of the 76,000 buildings in Hiroshima, 70,000 were damaged or destroyed. The estimated death toll eventual rose to 200,000 people—more than 60 percent of the population. And yet, the Japanese did not immediately surrender. To the civilian leadership of Japan, the bomb presented an opportunity for an honorable surrender, but the military leaders wouldn't hear of it.

The second atomic bomb was scheduled to be dropped on August 11, but the bomb date was pushed up to August 9 to avoid bad weather. Piloted by Charles Sweeney, *Bock's Car*, the B-29 carrying the plutonium bomb nicknamed "Fat Man," took off early on the morning of August 9. When the plane arrived at the primary target of Kokura, they found it hidden underneath a layer of thick smoke and haze, which made it impossible to make visual contact with the target. Captain Sweeney decided to try the secondary target, Nagasaki, but it, too, was shrouded in clouds. The decision was made to drop the bomb by using radar, but, at the last minute, a hole opened up in the clouds allowing the bombardier to take aim at a stadium located several miles from the planned aiming point. At 11:02, the plutonium bomb exploded 1,650 feet above the city of Nagasaki with an energy equivalent to 22,000 tons of TNT. The hills around Nagasaki shielded part of the city from the explosion, as a result, the damage was not as extreme as it had been at Hiroshima. Nevertheless, 140,000 people died at Nagasaki. With many admirals and generals still arguing against surrender, Emperor Hirohito of Japan stepped in and put an end to the slaughter. On August 15, 1945, the emperor, whose voice had never before been heard by the Japanese people, broadcast his message of surrender.

At war's end, the United States and the Soviet Union emerged as the dominant military superpowers on earth. But the United States had one weapon the Soviets didn't have: the atomic bomb. As long as the United States had a monopoly on the bomb, we had the upper hand. Aided by information gathered by spies in the Manhattan Project, most notably Los Alamos physicist Klaus Fuchs, the Soviets started work on an atomic bomb. Much to the surprise of U.S. experts, who had predicted it would take the Soviets many years to build the bomb, the Russians successfully tested an atomic bomb on August 29, 1949, only four years after the end of the war. In response, under the leadership of physicist Edward Teller, the United States began to pursue the hydrogen bomb, a bomb based on a process called nuclear fusion.

Nuclear fusion is the opposite of nuclear fission. In fission, a large nucleus is split into two smaller nuclei. In fusion, two small nuclei (for example, hydrogen) are combined or fused together to make a larger nucleus (for example, helium). In a fusion reaction, some original matter is converted into energy ($E = mc^2$ again). If enough nuclei fuse together, then enough energy is released to create a titanic explosion. But it's not easy to get two nuclei to fuse together. Why? Because nuclei contain positively charged protons, and like charges repel. To get these reluctant nuclei to fuse, they have to get very close together so that another force, called the strong nuclear force, can overcome the electrical force of repulsion. Getting the nuclei close together requires high densities and extreme temperatures on the order of a hundred million degrees. (Thus, a fusion weapon is sometimes referred to as a thermonuclear device.) At these temperatures, the nuclei are zipping around at incredible speeds, slamming into each other and getting close enough to allow the strong nuclear force to kick in.

How does a hydrogen bomb work? At one end of a hydrogen bomb sits an atomic bomb, at the other end, a supply of hydrogen. The atomic bomb is the trigger for a hydrogen bomb. When the atomic bomb is detonated, the explosion creates the high densities and temperatures required to make fusion happen. But why build a fusion-type bomb? Could it possibly be more powerful than an atomic bomb? The destructive power of a fusion bomb can be made to dwarf the power of an atomic bomb. The power of an atomic or fission-type bomb is limited by the critical mass, the minimum mass of fissionable material needed to sustain a chain reaction. A piece of uranium or plutonium with a mass greater than the critical mass cannot be placed in a fission bomb because stray neutrons may come in and initiate a chain reaction. With nuclear fusion, however, there is no chain reaction and therefore no critical mass problem. There is, in theory, no limit to the power of a fusion bomb. To make a bigger fusion bomb, just add hydrogen. The explosive power of atomic bombs is measured in thousands of tons (kilotons) of TNT; the power of thermonuclear weapons is often measured in millions of tons (megatons) of TNT—greater by a factor of one thousand. Modern nuclear weapons are thousands of times more powerful than the bomb dropped on Hiroshima. In fact, if one adds up the destructive power of all the conventional bombs dropped in World War II, the result is about two megatons. This is equivalent to a single, rather average nuclear warhead—a single weapon with the destructive force of World War II!

On November 1, 1952, at Eniwetok Atoll in the South Pacific, the United States tested the world's first thermonuclear device, code-named "Mike." It vaporized the island of Elugelab, left a crater 200 feet deep and more than a mile wide, and yielded an explosive energy equivalent to 10.4 megatons of TNT. Three years later on November 22, 1955, the Soviet Union successfully tested a hydrogen bomb. The United States had to have more nuclear weapons than the Soviets, and the Soviets had to keep pace with us. The arms race was on.

The next leg of the arms race involved the integration of the two new weapons systems developed during World War II: nuclear weapons and rockets. Instead of dropping an H-bomb from an airplane, a nuclear warhead could be blasted to its destination on top of a rocket. In August 1957, the Soviet Union successfully tested the world's first Intercontinental Ballistic Missile (ICBM); a Russian nuclear missile could reach the United States. Two months later, the Soviets shocked the world when it sent *Sputnik* into orbit, a demonstration of their superiority in space. The arms race had given birth to the space race. America quickly responded. In December 1957, the United States launched its first ICBM and in January 1958 sent *Explorer I* into space.

At the height of the arms race in the 1960s and early 1970s, the United States and the Soviet Union had thousands of nuclear weapons aimed at each other. Either country could now be totally annihilated at the push of a button. If the Soviets launched a nuclear attack against the United States, then the United States would retaliate and wipe out the U.S.S.R. If the United States attempted a first strike against the U.S.S.R, however, the Soviets would then destroy the United States. This nuclear stalemate became known as Mutually Assured Destruction (MAD). Although a nuclear holocaust was narrowly averted during the Cuban missile crisis in 1963, a nuclear exchange between the United States and the Soviet Union never happened. A combination of economic and political factors finally led to the dissolution of the Soviet Union in 1991 and the end of the cold war.

The sites described below are listed in approximate historical order with the Manhattan Projects sites followed by sites associated with the cold war. A visit to any of the Manhattan Project sites will be enhanced by viewing one of the many excellent movies and documentaries about the Manhattan Project. *The Day After Trinity* is a superb documentary about the life of J. Robert Oppenheimer. *Day One* is a historically accurate made-for-TV movie about the Manhattan Project. *Fat Man and Little Boy* is another good choice. Richard Rhoades wrote the definitive history of the development of the atomic bombs in *The Making of the Atomic Bomb*. A major effort is underway to clean up the old Manhattan Project sites. Many buildings are scheduled to be demolished, but a few will be preserved as national historic sites. In addition to the sites described below, several campus walking tours outlined in chapter 3 (especially the tours of Columbia University, the University of Chicago, and Berkeley) include sites related to the Manhattan Project.

## Manhattan Island, New York City

What better place to begin exploring Manhattan Project sites than Manhattan Island itself? The atomic bomb project was originally called the Laboratory for the Development of Substitute Materials, but Groves was afraid the name might arouse suspicion. Instead, Groves simply followed the standard procedure of naming any new unit after its geographical location. Because the headquarters were in Manhattan, the project became known as the Manhattan Engineering District. Over time, the name was shortened to the Manhattan Project. The role of Manhattan shrunk as secret cities in Los Alamos, Oak Ridge, and Hanford grew. Still, scientists and businesses based in Manhattan continued to play important roles throughout the project.

Let's begin our Manhattan tour at the southern tip of the island and work our way north. At 25 Broadway the Cunard Building housed the offices of a Belgian mining company, the African Metals Corporation. The Belgians had been warned that the Germans might be able to use uranium to make a new kind of super weapon. To keep the uranium out of German hands, the company had shipped 1,250 tons of high-grade uranium ore from its mine in the Belgian Congo to New York City. The ore was stored in 2,000 steel drums at Port Richmond on Staten Island. The Belgians had spent six months trying to inform the U.S. government about its presence. Somebody finally paid attention. On his first day as head of the Manhattan Project, General Groves sent an assistant to buy up all of the ore for a dollar per pound. The bill came to $2.5 million. African Metals eventually supplied the two-thirds of the uranium used throughout the Manhattan Project.

Proceed up Broadway to the Woolworth Building at 233 Broadway, directly west of City Hall. From 1913 until 1929, this sixty-story Gothic office tower was the tallest building in the world. In early 1942, the M. W. Kellogg Company, an engineering firm, was hired to design and build the gaseous diffusion plant in Tennessee. To preserve secrecy, the Kellogg Company created a front company, the Kellex Corporation. With a staff of several thousand, the Kellex Corporation occupied the eleventh, twelfth, and fourteenth floors of the building. One employee was physicist Klaus Fuchs, a Soviet spy who fed bomb secrets to the Russians. In these offices, experts from Columbia University and around the country met to determine the details of how to separate uranium isotopes by gaseous diffusion. Among the most difficult problems was devising a sturdy and effective porous barrier for the process. The problems solved here led to the success of the Oak Ridge plant.

Walk a little further up Broadway to Chambers Street. On the southeast corner at 270 Broadway is the twenty-seven-story Arthur Levitt State Office Building. Once the atomic bomb project had been approved, headquarter offices were established on the eighteenth floor of this nondescript building in June 1942. The building was chosen because the North Atlantic Division of the Army Corps of Engineers had their headquarters here, as did the Stone & Webster Engineering Corporation, a major contractor for the project. These offices served as headquarters for more than a year until Oak Ridge became the center of operations.

The next stop is West Twentieth Street, just off the West Side Highway near the Hudson River. Find the trio of buildings at 513–519, 521–527, and 529–535 West Twentieth Street. These are the former Baker and Williams warehouses where tons of uranium compounds were once stored. When uranium ore was purchased from the African Metals Corporation, it was shipped to a Canadian chemical refinery to be processed into uranium concentrates and then sent back to New York. The cargo ships moored along the west-side docks, where the uranium was loaded onto trucks for the short ride to the warehouses. Here, chemists tested each batch of uranium before it was distributed to other facilities. After the war, the warehouses continued to be used by the Atomic Energy Commission until the 1950s. The buildings underwent a thorough cleaning by the Department of Energy in the late 1980s and early 1990s.

Move now to the office building at 261 Fifth Avenue, on the southwestern corner of Fifth Avenue and East Twenty-ninth Street. When the original Manhattan Project offices were closed at 270 Broadway, they were relocated here to the new Madison Square Area Engineers Office. Here, a staff of several hundred on the twenty-second floor sought new sources for raw materials, arranged shipment to refineries, and oversaw the processing of the materials into a form that was ready for use. The project required a myriad of materials, including beryllium, chromium, fluorine, graphite, helium, lead, nickel, radium, and, of course, uranium ore.

Grab a subway to Grand Central Station, exit the building, and find 30 East Forty-second Street. This is the former headquarters of Union Carbide and Carbon Corporation, a company experienced in both exploring for and processing of minerals. You may be able to make out the company's name in faint lettering on the building's north side. The Manhattan Project enlisted the help of Union Carbide in discovering new sources for uranium ore. On the eighteenth floor the company set up a subsidiary company, the Union Mines Development Corporation, to lead this effort. Union Carbide was eventually chosen to operate the giant K-25 gaseous diffusion plant at Oak Ridge. General Groves, Columbia scientists, and experts from the Kellex Corporation met here in early 1943 to give the company its first briefing about the plant.

The next stop on our tour is an apartment building at 155 Riverside Drive at West Eighty-eighth Street, the boyhood home of J. Robert Oppenheimer, the scientific director of the Manhattan Project. His father was a successful textile importer, his mother, a teacher and painter. The family lived on the eleventh floor overlooking the Hudson River. The apartment was decorated with original artwork by Picasso, Rembrandt, Renoir, and Van Gogh. For more on Oppenheimer, refer to the entry in chapter 1.

Continue north on Riverside Drive until you find yourself on the block between 105th and 106th streets. Here, in front of the New York Buddhist Church is a statue of Shinran Shonin, a Japanese Buddhist monk who lived during the twelfth and thirteenth centuries. He is depicted in his missionary travel robes, with a wooden staff clasped in his right hand and his head crowned with an umbrellalike peasant hat. The statue was brought to New York from Hiroshima, where it stood a little more than a mile away from the center of the atomic bomb blast. Somehow, it survived. A plaque declares the statue is "a testimonial to the atomic bomb devastation and a symbol of lasting hope for world peace."

Jump back on the subway and take it to 116th Street, which is Columbia University. Facing the famous domed Low Library, walk around it to the left. Straight ahead is Pupin Hall, home of the Columbia University Physics Department. Some initial experiments that led to the Manhattan Project took place in the basement floors of this building. Here Enrico Fermi replicated the German experiment in which nuclear fission was first demonstrated. On the eighth floor is a memorial room commemorating the life and work of Nobel Laureate and Manhattan Project scientist I. I. Rabi. Gaseous diffusion research was done in nearby Havemeyer Hall. Consult the entry on Columbia University in chapter 3 for more details.

Just east of the campus at 420 West 116th Street is a nine-story, tan brick building. This was once the King's Crown Hotel where Leo Szilard stayed while he worked at Columbia University. Though less well known than some scientific stars of the Manhattan Project, Szilard was a major figure in the atomic bomb saga. As I mentioned in the introduction to this chapter, Szilard came up with the idea of a chain reaction, and at his urging Einstein signed the letter to Roosevelt warning of the possibility of a German weapon. Szilard had many clashes with General Groves, who once described Szilard as "the kind of man that any employer would have fired as a troublemaker." Near the end of the war, Szilard circulated a petition that argued against using the bomb against the Japanese. The petition never made it to President Truman.

Our final stop is the three-story Nash Garage Building at 3310 Broadway, between West 133rd and 134th streets. Before World War II, this car dealership sold the popular Nash Ambassador 600. During the war, all nonessential automotive production was halted. Columbia University acquired the space to produce barriers and test equipment for use in the gaseous diffusion plant in Oak Ridge. Today, the ground floor of the building houses a department store while the upper two stories are used as a parking garage.

## Nuclear Energy Sculpture, Chicago, Illinois

To make a plutonium bomb, one must first make plutonium, an element that doesn't exist in nature. To make plutonium, you need a nuclear reactor. The world's first nuclear reactor was built not to generate electricity—that had to wait until after the war. Instead, scientists built the reactor to demonstrate two phenomena: first, a chain reaction is possible; second, the chain reaction could produce plutonium. The task of building the first reactor fell on the capable shoulders of Italian Nobel Prize winner Enrico Fermi, who was unusual among physicists because he was equally adept at both theoretical and experimental work.

The reactor Fermi and his team built, known as Chicago Pile #1 or CP-1, was roughly in the shape of a beehive, 25 feet wide and 20 feet high, flat and rectangular toward the bottom and rounded toward the top. It consisted of a complex matrix of fifty-seven alternating layers of graphite and uranium held together by a wooden frame and covered on all sides except one by material from a dark grey Goodyear balloon. It took less than a month to build at a cost of about one million dollars. Fermi and company used blocks of graphite, a form of carbon found in pencils, to slow down or "moderate" the speed of the neutrons released by the fissioning of the uranium nuclei. Slow-moving neutrons have a better chance of causing a fission that a fast-moving neutron, just as a slow-moving golf ball has a better chance of dropping in the hole than a ball that whizzes by. The more time the ball spends around the hole, the better chance it has of falling in; similarly, the more time a neutron spends around a uranium nucleus, the better chance it has of hitting the nucleus and causing it to split.

Originally, scientists planned to piece together the reactor in a building twenty miles southwest of Chicago in the Argonne forest, but the workers who were supposed to construct the building went on strike. Fermi then suggested to his boss, Arthur Compton, physics Nobel Laureate and administrator of the plutonium project, that the squash courts on the University of Chicago campus would be a suitable alternative because his team had already done some preliminary experiments there.

But wouldn't it be a tad dangerous to build the first experimental nuclear reactor in the middle of what was then the second largest city in the country? The safety issue was carefully considered. First, there was never any risk of a nuclear explosion. The laws of physics would prevent that. There was a risk of what is now known as a meltdown, which happens when the reactor overheats and the fuel literally melts. This was to be a controlled nuclear chain reaction. How do you control it? By moving cadmium control rods into and out of the reactor. Cadmium is an element that soaks up neutrons. Because the neutrons make the chain reaction happen, if you control the neutrons, then you control the reactor. By pulling the control rods out of the reactor, the rods release the neutrons to start the chain reaction. By pushing the control rods into the reactor, the rods absorb the neutrons, thereby snuffing out the chain reaction. Fermi had several control rods poised and at the ready. Following Fermi's instructions an assistant manually operated one control rod. Another safety control rod, operated automatically, was set to be released if the intensity of the neutrons reached a preset limit. Yet another control rod was suspended above the reactor on a rope, where stood another nervous young physicist, armed with an ax and ready, if called upon, to cut the rope and let the rod plunge into the heart of the reactor. A final safety precaution consisted of a "suicide squad" of three scientists holding jugs of cadmium sulfate solution above the reactor; they could pour the solution on the pile and dowse the chain reaction. Some experiments showed that a small fraction of the neutrons released in the fission process would not come out immediately, but would be delayed for a few seconds; these ultimately convinced Compton to give the go-ahead. This delay allowed Fermi, who was operating the reactor just barely above the level needed to sustain the reaction, time to react. Compton's decision indicates the supreme confidence he had in Fermi's engineering skill. The scientists informed neither the president of the University of Chicago nor the mayor of Chicago of the plan.

On December 2, 1942, the experiment began around mid-morning. Fermi ordered all but one control rod removed from the pile. After checking some measurements, he ordered the last control rod withdrawn about halfway. Fermi made some calculations. Then the rod was withdrawn another six inches, followed by more calculations and checking. This pattern continued throughout the morning. After one six-inch withdrawal of the rod, a loud crash was heard. The automatic control rod had been released. The safety point at which the rod was released had been set too low. Fermi announced it was time for lunch, and the control rods were placed back into the reactor. At around two in the afternoon, Fermi continued the experiment. A crowd of forty-two people, mostly physicists who had worked on the pile, gathered on the balcony and the squash court. Leo Szilard was there along with Arthur Compton. The process of withdrawing the last remaining control rod in careful increments continued. At 3:25 P.M., Fermi ordered the control rod out about a foot, the reaction became self-sustaining, and the nuclear age was born.

### Visiting Information

The twelve-foot bronze sculpture titled *Nuclear Energy* was erected on the site of Stagg Field where the squash courts were located underneath the bleachers. Some interpret the abstract sculpture as representing a mushroom cloud. To others, it resembles a human skull. Sculptor Henry Moore said that he hoped those viewing the piece would move around it, look through the open spaces, and have the feeling of being inside a cathedral. The sculpture is inscribed with the following statement: "On December 2, 1942, man achieved here the first self-sustaining chain reaction and thereby initiated the controlled release of nuclear energy." The sculpture is on the east side of Ellis Avenue between 56th and 57th streets on the campus of the University of Chicago. (See the entry on the University of Chicago in chapter 3 for more scientific sites.)

## Oak Ridge, Tennessee

In September 1942, as one of his first official acts, General Groves chose a 59,000-acre site in eastern Tennessee to build the facilities needed for the Manhattan Project. The selection was based on several factors. Geologically, the area consisted of a series of parallel ridge valleys aligned in a southwesterly direction. The various facilities could be built in different valleys so that, if there were an explosion, then the other facilities would be shielded by the hills. There was also an available labor pool in nearby Knoxville, and a new TVA dam could provide electrical power. According to some sources, the decision to locate the site in Tennessee was influenced by Tennessee Senator K. D. McKellar, who just happened to be the chairman of the Senate Appropriations Committee. When President Roosevelt asked McKellar to quietly approve the funds needed to build the top-secret facility (60 percent of the entire Manhattan Project budget was spent at Oak Ridge), the chairman replied: "Yes, it can be done, and Mr. President, just where in Tennessee are we going to locate that thing?" The 3,000 residents of five small communities were given only a few weeks to vacate their property. Oak Ridge's population quickly ballooned to 75,000: 800 buses carried 120,000 passengers daily; 17 eating establishments served 40,000 daily meals; 35,000 housing units appeared; dormitories housed 15,000; and schools educated 11,000 students. Yet, the city was truly secret: it did not appear on any map and was surrounded by barbed wire, guards were posted at all entrances to the city, and residents were required to wear badges outside their homes. The city was not open to the public until 1949, and the current population is down to 28,000.

What was done here at this "Secret City"? One of the most difficult obstacles faced by the Manhattan Project scientists was this: only 0.7 percent of natural uranium dug out of the ground is U-235, the fissionable isotope of uranium. The rest is nonfissionable U-238. Therefore, to make a uranium atomic bomb, the U-235 must be separated from the U-238. This is not an easy task because both U-235 and U-238 are, of course, the same chemical element, with the same chemical properties and behavior. The only difference between U-235 and U-238 is that U-238 has three more neutrons, which makes it slightly heavier. To separate the isotopes, one must take advantage of this tiny difference in mass. The process of increasing the percentage of U-235 in a sample of natural uranium is called uranium enrichment. The primary task of the Oak Ridge facilities was to produce highly enriched uranium for use in the atomic bombs. To accomplish this, Oak Ridge used two main methods of isotopic separation: electromagnetic separation and gaseous diffusion. Electromagnetic separation made use of the fact that when charged particles move through a magnetic field, a force exerted on the particles causes them to curve. The radius of the curved path is directly proportional to the mass of the particle: the bigger the mass, the larger the radius of curvature. In electromagnetic separation, natural uranium is combined with chlorine to form uranium tetrachloride, which is ionized and injected into a vacuum chamber that contains a magnetic field pointing perpendicular to the direction of motion of the particles. The uranium tetrachloride molecules with the U-235 executed a curve with a smaller radius than the molecules with the U-238. The machines that produce this electromagnetic separation are called calutrons (a variation on a device physicists call a "mass spectrometer") because they were invented at the University of California. More than a thousand calutrons were built and housed in the Y-12 alpha and beta buildings at Oak Ridge. The alpha calutrons enriched the uranium up to 15 percent. This enriched uranium was then fed into the beta calutrons, which produced weapons-grade highly enriched uranium. The calutrons needed electromagnets to produce a magnetic field. Normally, copper wires would be used to make the coils for the electromagnets, but the war effort had created a shortage of copper. Amazingly, the U.S. Treasury Department melted down its supply of silver, 14,000 tons worth, so that silver wires could be used in the calutrons!

At its peak, Y-12 employed 22,000 workers, many of whom were women. These "calutron girls" were trained to sit and watch meters on the control panels and make adjustments if the needles moved too far in one direction or the other. Today, the Y-12 National Security Facility, an entity separate from ORNL, is the nation's Fort Knox for highly enriched uranium. The facility, a world leader in uranium technology, is responsible for inspecting, refurbishing, dismantling, and nondestructively testing the nation's nuclear arsenal.

The second method of isotopic separation is known as gaseous diffusion. In this process, uranium is combined with fluorine to form a gas called uranium hexafluoride or $UF_6$. The gas is pumped against a porous barrier containing millions of submicroscopic holes in every square inch. The idea is that the lighter molecules with the U-235 pass or diffuse through the pores more rapidly than the heavier molecules with the U-238. As with electromagnetic separation, this process must be repeated thousands of times to adequately enrich the uranium for use in a bomb. At Oak Ridge, the diffusion process was housed in the titanic K-25 building. Standing four stories tall, a half-mile long, 1,000 feet wide, and covering nearly 43 acres, it remains one of the largest buildings in the world. Today, plans are underway at the old K-25 site, called the East Tennessee Technology Park, to encourage reindustrialization by the private sector.

The Oak Ridge facilities also housed the Graphite Reactor, a scaled-up version of Fermi's Chicago reactor. This reactor was built to make sure that plutonium could be produced from uranium in a reactor before investing in large reactors at Hanford, Washington. Before dawn on the morning of November 4, 1943, Enrico Fermi and Arthur Compton were summoned from the guest house to witness the reactor achieve criticality. A few months later, it was producing some of the world's first plutonium. The reactor was subsequently used to find out which construction materials absorbed the fewest neutrons so that scientists could build future reactors of substances that minimized the depletion of neutrons. The optimal ratio of graphite to

Oak Ridge's giant K-25 Building, housing the machinery for the gaseous diffusion process that separated uranium isotopes.

uranium was another important issue resolved by the Graphite Reactor. After the war, researchers used the reactor to generate electricity from nuclear energy—the idea of a light water reactor originated here—and to educate students on nuclear technology. Admiral Hyman Rickover, father of the nuclear navy, was an illustrious graduate of the school. Perhaps the most significant postwar work was producing and distributing radioactive isotopes or "radioisotopes." Radioisotopes are produced when the neutrons in a nuclear reactor bombard stable nuclei. Some neutrons stick in the nucleus, which makes it unstable (radioactive). In the first year of operation, more than a thousand shipments of radioisotopes were made. By 1950, the number had increased to nearly 20,000. These radioactive isotopes have been used for medical diagnosis, treatment, and imaging, for tracing the action of fertilizers, and for deciphering the genetic code. The Graphite Reactor was shut down in 1963 and production shifted to the Oak Ridge Research Reactor which operated until 1987. Today, most radioisotopes are foreignmade, but the High Flux Isotope Reactor continues to make isotopes for special purposes. The Graphite Reactor, the oldest existing nuclear reactor in the world, has been designated as a National Historic Landmark.

The area around the X-10 Graphite Reactor is now the heart of Oak Ridge National Laboratory (ORNL). Today, the lab's six primary missions are in the areas of neutron science, computational science, materials science (including nanoscale research), microbial biology and proteomics, energy

Courtesy Ed Wescott

The face of the Graphite Reactor at Oak Ridge.

technology, and security technologies and nuclear nonproliferation programs. The lab maintains eight user facilities, including the Mouse Genetics Research Facility, home to 250,000 mice used for genetics and genomics research. The newest facility is the Spallation Neutron Source (SNS), a $1.4-billion facility for the neutron analysis of materials. The lab employs more than 4,000 people, including 1500 scientists and engineers, and operates on an annual budget of $1.1 billion.

## Visiting Information

The best place to start your Oak Ridge odyssey is at the American Museum of Science and Energy (AMSE). Here, you can learn all about this "Secret City," sign up for the bus tour, and pick up maps and brochures that include scientific points of interest relating to the Manhattan Project. Movies about Oak Ridge can be viewed in the auditorium, and their presentations are announced throughout the day. The museum's main exhibit hall on the first floor is called the "The Oak Ridge Story." This hall traces the history of Oak Ridge from the acquisition of the land to the dropping of the bomb. Notable artifacts include the guts of a calutron complete with its control panel, models of the X-10 reactor along with a graph showing the blip in neutron intensity as the reactor goes critical, a pilot model of a gaseous diffusion chamber, a reproduction of the navigator's log for the Hiroshima flight, and samples of Trinitite. Upstairs are several more exhibit halls, including one called "The World of the Atom" with displays on all aspects of nuclear energy. One of the more interesting panels describes a new method of isotopic separation called "Atomic Laser Vapor Isotope Separation (ALVIS)." Laugh out loud at the "Revigator," a quack medical device that supposedly uses radiation to cure whatever ails you. The Y-12 National Security Complex hall holds mock-ups of several nuclear weapons including the Mark 28 and the B-83, respectively the oldest and newest thermonuclear bombs in the U.S. Arsenal. Don't miss the model of "Little Boy," the Hiroshima bomb that used Oak Ridge's uranium. Outside the museum near the entrance is a gray gaseous diffusion chamber. The museum is open daily except New Year's Day, Thanksgiving, and Christmas. Museum hours are Monday through Saturday, 9:00 A.M. until 5:00 P.M., and Sunday, 1:00 P.M. until 5:00 P.M. Admission is $5 for adults, $3 for children age 6 to 17, and $4 for senior citizens age 65 and older. Children age 5 and under are free.

After visiting the museum, you're ready to view the sites via bus, train, and/or car. Bus tours leave from the AMSE at noon on weekdays during the summer months. Tickets for the bus tour are included in your admission to the AMSE, but the limited number (about twenty-four) are distributed on a first-come, first-served basis on the day of the tour. Get to the museum early to insure a seat on the bus. The two-and-a half-hour tours stop at three or four sites, including a couple of sites within ORNL that are not publicly accessible. The first stop is the Visitor's Center at the Y-12 National Security Complex. Y-12, the first site developed at Oak Ridge, separated isotopes of uranium using calutrons. After a brief video and a presentation on the town of New Hope, one of five small communities taken over by the Manhattan Project, you can walk around the room to inspect various pieces of equipment used at Y-12, including lathes, scales, electrical meters, and a weld inspection kit. The next stop is the Spallation Neutron Source inside ORNL. Unfortunately, you don't actually get to see anything except a big diagram in the lobby—very disappointing. (The SNS stop is only included on the Friday bus tours.) The highlight of the tour is the Graphite Reactor. When I stood in front of the face of the reactor, I fully experienced its enormous size. Then it's on to the control room and an experimental area where materials were inserted into the reactor. The final stop is the K-25 scenic overlook where you can see part of the gigantic U-shaped, K-25 building, where the gaseous diffusion method of isotopic separation was used. A pavilion here provides a video and photographs. A photo above the display identifies the buildings. To be honest, from here the view of the K-25 building is not very good. A much better view can be had by taking the Perimeter Road that loops around the site. The two half-mile-long legs of the K-25 building are scheduled for cleanup and demolition. An effort is underway to preserve the northern end of the building and convert it into museum.

Another way to experience Oak Ridge is on the "Secret City Scenic Excursion Train." These one-hour, narrated train rides take you to the K-25 site only. The rides cost $15 for adults and $10 for children and are usually scheduled on the first and third Saturdays of each month. Check the website below for a complete schedule. Reservations, although not required, are recommended during the busy summer months. Credit cards are not accepted for payment.

You can get to all major historical sites, except the graphite reactor, on your own in your car. (You can make arrangements for a private tour of the graphite reactor by calling the number below.) An "Oak Ridge Driving Tour" brochure is available at the AMSE and the Visitor Center. An audio

CD that accompanies the driving tour is also available for purchase. The driving tour includes a number of sites unrelated to the Manhattan Project, but a few may be of interest to the scientific traveler. The original guest house, the Alexander Inn, was the only army hotel within the gates of the city. Nearby, the Chapel on the Hill held multiple services for different denominations during the war years. The International Friendship Bell, cast by a Japanese bell-maker, is the first monument dedicated to peace between a Manhattan Project city and Japan. An original military guard house is also on the tour. Instead of the driving tour, another brochure, "Unlock the Secrets of America's Secret City," circles and describes points of interest on a map.

In June, Oak Ridge has a "Secret City Festival" celebrating the end of World War II and the city's heritage. Activities include the largest World War II reenactment in the South, and special Manhattan Project tours include areas not usually open to the public. One of these tours is of the K-12 facility where you might see Building 9731, the oldest process building at Y-12 and the Beta-3 building, which still holds calutrons in operable condition. After the clean-up, these will be the only remaining K-12 buildings.

| | |
|---|---|
| Websites: | www.OakRidgeVisitor.com (visitor information) |
| | www.amse.org (museum) |
| | www.southernappalachia.railway.museum (excursion train) |
| | www.secretcityfestival.com |
| Telephone: | 800–887–3429 (visitor information) |
| | 865–576–3200 (museum) |
| | 865–574–4165 (Graphite Reactor tours) |
| | 865–241–2140 (excursion train) |

## Hanford Site, Hanford, Washington

After the first experimental reactor was built in Chicago by Enrico Fermi, a scaled-up graphite reactor was constructed at Oak Ridge to ensure that plutonium could be produced by this method. Following the success of the Oak Ridge Graphite Reactor, full-scale production reactors were built here. The reactors at Hanford site produced the plutonium that was used in the device tested at Trinity site in July 1945 and the "Fat Man" bomb dropped on the city of Nagasaki, Japan, in August.

How is plutonium produced in a nuclear reactor? The uranium fuel used in a reactor is not pure U-235, but rather a mix of U-235 and the more abundant U-238. When a U-235 nucleus fissions, it releases neutrons. Some of those neutrons will be absorbed by the U-238, making U-239. One neutron in the U-239 will change or "decay" into a proton and an electron; the proton remains in the nucleus while the electron is spat out. This process is called beta decay. Now, recall from your high school chemistry class that the chemical identity of a nucleus is determined solely by the number of protons in the nucleus. Because a proton has been added to the U-239 nucleus, it is no longer uranium, but the next element on the periodic table, neptunium. The Np-239 nucleus repeats the beta-decay process, and yet another proton is added to the nucleus. This results in element number 94, plutonium. The amount of plutonium generated depends on the length and power level of the reactor's operation. When the fuel is removed from the reactor, the plutonium can be separated from the uranium by ordinary chemical methods because they are two distinct chemical elements. This technique avoids the arduous task of isotopic separation. A nuclear reactor can be used not just to generate electricity but to make plutonium. That's why the United States gets nervous when rogue countries like Iran and North Korea have access to nuclear technology.

Owing to safety considerations, General Groves decided not to locate the plutonium production facilities at Oak Ridge. The city of Knoxville was close, and Groves worried that, in the highly unlikely event of an explosion, winds might carry radioactive materials over the city, endangering the population, contaminating the uranium separation facilities, and revealing the secret of the Manhattan Project. A suitable site must be in a remote location with an abundant water supply and plenty of electricity. Groves found the Hanford site along the Columbia River in south central Washington to be ideal in all respects and bought the 640-square-mile tract of land for $5.1 million dollars in January 1943. The first reactor, B-reactor, produced its first plutonium in November 1944, and the first plutonium shipment reached Los Alamos in February of 1945.

After the war, the Hanford reactors were used to investigate the properties and behavior of nuclear reactors and to improve their operating efficiencies. The B-reactor produced tritium for the hydrogen bomb. Most Hanford reactors were mothballed during the 1960s, but radioactive waste remains at the site. Today, Hanford is one of the world's largest environmental clean-up sites. Thousands of people are working to consolidate the waste and clean up contaminated buildings and soil. Administrators estimate that the clean-up efforts will last until 2030.

## Visiting Information

Visiting Hanford site ain't easy! Check the Hanford website to see when the next session of public tours are scheduled. You must register for the tours online, and you may request up to two tour slots for your party. The tour slots are filled on a first-come, first-served basis. The tour is very popular so be sure to register as soon as the online registration becomes available for the session you are interested in. To take the tour, you must be a U.S. citizen at least sixteen years old. You will be asked to present a photo ID (for example, a driver's license) on the day of the tour, and you must wear clothing suitable for an industrial environment. No shorts, sleeveless shirts, or open-toed shoes. The free tours last about five hours.

The tours follow a route designed to trace the steps used to produce plutonium. The route includes the 300 Area, where the fuel was manufactured, the original Hanford town site, the production reactors along the Columbia River, a walking tour of the historic B-reactor, and the 200 Area where the nuclear fuel was processed to extract the plutonium. The website offers a virtual tour of Hanford and has several online videos about Hanford site. The tours begin and end at the Volpentest HAMMER

> Website:  www.hanford.gov
> Telephone:  509–376–3375

Training & Education Center, located at 2890 Horn Rapids Road in Richland, Washington. Richland is about a three-hour drive from Seattle.

## Los Alamos, New Mexico

In 1942, General Groves needed a location for the bomb design lab—a place where the scientists could talk and argue freely amongst themselves, but be separated from the eyes and ears of the general population. Groves claimed he wanted an isolated place for safety's sake; if anything went wrong at the lab, he didn't want nearby communities affected. Clearly, the isolation also enhanced the security of the top-secret project. Groves assigned the job of finding the right location for the lab to Major John H. Dudley and gave him the following selection criteria: some existing facilities and room for 265 people, west of the Mississippi and at least two hundred miles from the U.S. border, and surrounded by hills so that a fence could be installed and easily guarded. Dudley's first choice, Oak City, Utah, was rejected because too many families would have to be displaced and too much farmland would have to be rendered unproductive. The second choice was Jemez Springs, New Mexico, but when Oppenheimer and Groves visited the site, they didn't like it. The site lay in a canyon, and Oppenheimer wanted a lab with a view. As a youth, Oppenheimer had spent a summer in the area and knew it well. He told Groves about a boys' school on a nearby mesa that might be a suitable site. The name of the school was Los Alamos, the Spanish term for the cottonwood trees that lined a stream on the mesa. A fence could be strung along the rim of the mesa, satisfying Groves, and Oppenheimer would get his view. And so it was that Los Alamos became the place where the world's first nuclear weapons were designed and built.

Today, the Los Alamos National Laboratory (LANL) is spread over thirty-eight square miles and employs more than 11,000 people. The lab's main mission is referred to as "stockpile stewardship." In other words, the lab is responsible for ensuring that our aging nuclear weapons arsenal will work should, heaven forbid, they be needed. In fact, the United States has not built any new nuclear weapons since 1992. International treaties forbid the United States from testing the weapons directly by actually detonating them. As a result, the lab must rely on simulations using supercomputers and nonnuclear explosives and on applying physics and chemistry to the various components of the weapons to make science-based predictions on how they will behave.

There are no public tours of LANL, although you are free to drive around the grounds. In 2000, a devastating wildfire destroyed more than two hundred homes in Los Alamos, forced the evacuation of 20,000 residents, and got dangerously close to the lab. If you look at the hillsides to the north, you can see some fire damage.

Only the Otowi building, which has a cafeteria and a series of five memorial markers near the entrance, is open to the public. LANL operates the Bradbury Science Museum, a liaison between the lab and the public. The museum is named in honor of Norris Bradbury, the lab's second director from 1945 to 1970 (not Ray Bradbury, the science fiction writer). Start your museum visit with the history gallery, which tells the story of the Manhattan Project. Try to catch the sixteen-minute film *The Town That Never Was* in the adjacent auditorium. Listen to the announcements for the next show time. Proceed back out into the lobby and watch the short video on *The Lab Today*, which gives you an overview of the current research conducted at LANL. The defense and research galleries showcase bomb casings of "Fat Man" and "Little Boy" and a dummy unit of a B83, the most powerful weapon in the U.S. nuclear arsenal. In the theater, watch *Mission: Stockpile Stewardship*, a film that describes how the lab certifies the readiness

Jeffrey M. Frank / Shutterstock

A mock-up of the plutonium bomb, nicknamed "Fat Man," on display at the Bradbury Science Museum at Los Alamos.

of our nuclear weapons. Much to its credit, the museum has a public forum space set aside for displays by groups, independent from the lab, that offer diverse opinions on nuclear weapons and their use. A book is available for individual guests to express their own opinions on nuclear issues. The Bradbury Science Museum is free and open from 1:00 P.M. to 5:00 P.M. on Sunday and Monday and 10:00 A.M. to 5:00 P.M. the rest of the week. The Otawi Station Bookstore, next to the museum, has a variety of gifts and souvenirs.

Buffalo Tours offer one-and-a-half-hour van tours of the Los Alamos area for $15 per person. You can purchase tickets at the bookstore where the tours depart at 1:30 P.M. from March through October. They also advertise a longer (and more expensive) "Off the Hill Manhattan Project" tour, which includes the Lamy train station and various sites in Sante Fe related to the Manhattan Project. If you want to explore on your own, you can purchase a detailed map of the Los Alamos area found in the Visitors Guide, available in the bookstore or at the chamber of commerce.

Another "must see" site is the Los Alamos Historical Museum, housed in the ranch school guest cottage where General Groves stayed when he was at the lab. This small museum focuses on the history of the Los Alamos area including the natural history and geology, the Native American inhabitants, the Los Alamos Ranch School, and, of course, the Manhattan Project. Here, you can see a sample of Trinitite, a glasslike rock formed when the heat of the atomic bomb test melted and fused the desert sand. Nearby is a piece of thick green glass used as eye protection at Trinity along with slide rules and textbooks used during the project. Panoramic 360° photos of Hiroshima and Nagasaki give the visitor an idea of the extent of the devastation. The pictures are supplemented by artifacts including a pair of sunglasses used aboard the *Enola Gay* and a partially melted pocket watch found a quarter of a mile from the center of the Hiroshima blast. An unusual assortment of cold war era artifacts includes a lamp in the shape of a bomb. To the left of the entrance is a little bookstore and gift shop that's worth a quick look. The Historical Museum is free and open Monday through Friday from 10:00 A.M. to 4:00 P.M. and from 1:00 P.M. to 4:00 P.M. on Saturday and Sunday. The Historical Museum is a short five-minute walk from the Bradbury Museum. Just stroll up Central Avenue to Twentieth Street and look to your right. The museum is behind the Fuller Lodge.

At the museum, be sure to pick up a guide for the historical society's self-guided thirteen-stop walking tour of Los Alamos. This is an absolute "must do" because it takes you to a few surviving buildings from the Manhattan Project. The walk includes the Baker House, just west of the museum, where Sir James Chadwick, discoverer of the neutron, stayed while serving as head of the British mission to the Manhattan Project. The Ice House Memorial marks the place where the nuclear components of the first atomic bomb were assembled before transporting them to Trinity site. Further along, you'll come to the power house where the leading Manhattan Project explosives expert, George Kistiakowsky, lived. The highlight of the walking tour is a line of houses dubbed "bathtub row" because back in the early days of the project they were the only houses that had bathtubs. Enrico Fermi lived at 1300 Twentieth Street, a house that was later occupied by Norris Bradbury. The house at 1967 Peach Street, at the corner of Twentieth and Peach Streets, was home to Oppenheimer and his family. Other notable homes not on the walking tour include Edward Teller's small single-story house at 1016 Forty-ninth Street on the far west side of the North Mesa. The Unitarian Church on the northwest corner of Sage Loop and Fifteenth Street is the only remaining bachelor dormitory, and the second floor rooms remain in their original condition. Richard Feynman lived here while his sick wife rested in an Albuquerque hospital.

The most unusual place to visit in Los Alamos must be the Black Hole (a.k.a. the Los Alamos Sales Company), a military surplus store holding thousands of items from the LANL and elsewhere. The store claims it houses "the world's most diverse stock of used scientific equipment, electronics, lab supplies, nuclear by-products, surplus items, and materials." This legendary establishment is owned and operated by Ed Grothus, a former employee of the LANL. Ed, who worked at the lab for twenty years as a machinist and technician, has evidently had a change of heart regarding nuclear weapons. He is now an avid antinuclear weapons activist. A sign above the door reads: "one bomb is too much . . . no one is safe unless everyone is safe . . . don't throw anything out . . . welcome to the black hole." The eccentric and colorful Ed has attained celebrity status and has been the subject of three documentary films. The Black Hole is located at 4015 Arkansas Avenue.

Los Alamos is about a forty-minute drive from Santa Fe. If you plan to stop in Santa Fe, there are a couple of Manhattan Project sites you may want to visit. When new Los Alamos personnel arrived, they

Websites: www.lanl.gov/museum
(Bradbury Science Museum)
www.losalamoshistory.org
(Los Alamos Historical Museum)
http://buffalotours.home.att.net (tours)
http://members.aol.com/blkholela/home
(Black Hole exhibit)

Telephone: 505–667–4444 (science museum)
505–662–6272 (historical museum)
505–662–3965 (buffalo tours)
505–662–5053 (Black Hole)

checked in at the U.S. Engineer Office No. 3 situated in a courtyard at 109 East Palace Avenue. A plaque on the north wall relays the history of the office. Scientists and other visitors who lacked clearance to enter Los Alamos often got a room at the La Fonda Hotel on San Francisco Street. Klaus Fuchs, the infamous spy who passed bomb secrets to the Soviets, occasionally met his contact on the Delgado Street Bridge.

# Trinity Site National Historic Monument, Near Alamogordo, New Mexico

On July 16, 1945, at 5:29:45 A.M., the world's first nuclear weapon was detonated atop a 100-foot-tall steel tower at this barren and remote site in the New Mexican desert. The explosion, produced by the splitting of billions of plutonium nuclei, released an amount of energy equivalent to nearly 20,000 tons of TNT and created a crater 2,400 feet wide and 10 feet deep. A brilliant flash of light brighter than a dozen suns was seen throughout New Mexico and in parts of Arizona, Texas, and Mexico. The temperature near the center of the blast was approximately 10 million degrees, melting and fusing the desert sand into a jade-colored glass, later known as "Trinitite." Within minutes, a multicolored mushroom shaped cloud had roiled and gurgled its way seven miles into the predawn sky. The accompanying shock wave shattered windows 120 miles away.

J. Robert Oppenheimer, scientific head of the Manhattan Project, described his reaction in this way:

> We waited until the blast had passed, walked out of the shelter and then it was extremely solemn. We knew the world would not be the same. A few people laughed, a few peopled cried. Most people were silent. I remembered the line from the Hindu scripture, the Bhagavad-Gita: Vishnu is trying to persuade the Prince that he should do his duty and to impress him he takes on his multi-armed form and says, "Now I am become Death, the destroyer of worlds." I suppose we all thought that, one way or another.

It would be difficult to overstate the historic importance of this event and this place. Think of it: the world's geopolitical landscape during the last half of the twentieth century was dominated by the specter of a nuclear war between the United States and the Soviet Union. During a brief interlude of a few years following the dissolution of the old Soviet Union, it looked like we might not have to worry about nuclear weapons anymore. September 11, 2001, changed all that. We are now faced with the sobering possibility that a rogue nation or a terrorist organization might use these weapons, weapons that were first tested here, in this place, on this spot.

Trinity site was selected for the atomic bomb test on the basis of several criteria:

1. It was isolated to help insure secrecy and safety;
2. Good weather was the norm;
3. There was virtually no one living anywhere near the site;
4. It was far enough away to avoid an obvious connection with Los Alamos; and
5. It was near enough to make travel convenient.

Interestingly enough, this site was not the first choice of General Groves. He had wanted to use a military training area in southern California but discovered that he would have to get permission from General George S. Patton, a man who Groves once described as "the most disagreeable man I

The marker at Trinity Site in New Mexico, where the first atomic bomb was tested in July 1945.

had ever met." Rather than having to deal with Patton, Groves settled for his second choice.

What is the origin and significance, if any, of the code-name "Trinity"? This is not entirely clear. According to the most popular explanation, Oppenheimer based the name Trinity on a devotional poem by the six-teenth-century English poet John Donne. The poem began with the line "Better my heart, three person'd God." Another Donne poem suggested to Oppenheimer the idea of death and resurrection, a concept that Oppen-heimer saw reflected in the fact that although the bomb would most cer-tainly cause death on a massive scale, it might also lead to the end of the war and the redemption of mankind.

## Visiting Information

A visitor to Trinity site can stand at Ground Zero, which is marked by a mod-est monument built of black lava rock. A plaque on the monument bears this inscription: "Trinity Site Where the World's First Nuclear Device Was Exploded on July 16, 1945." The crater has been filled in, and most of the Trinitite has been removed. Visitors can also see the McDonald ranch house where the plutonium core for the bomb was assembled.

Photography is allowed at Trinity Site but is forbidden anywhere else on the range. Portable toilet facilities are provided on the site, and hot dogs and soft drinks are sold in the parking lot. Be sure to check the White Sands Mis-sile Range website for up-to-date information or call the public affairs office at (505) 678–1134 Ext. 1700.

Finally, a word about radiation. The radiation levels in the fenced area around ground zero are low, only about ten times higher than the region's natural background radiation. A one-hour visit to ground zero results in an exposure of one-half to one milliroentgen of radiation. For comparison, fly-ing coast to coast in a commercial jetliner results in an exposure of between three and five milliroentgens.

Visiting Trinity Site requires determination. Because the site is located on an active firing range, it is normally open to the public only twice a year: the first Saturdays in April and October. On tour days, visitors can enter White Sands Missile Range at the Stallion Range Center located five miles south of highway 380. The turnoff is twelve miles east of San Antonio and fifty-three miles west of Carrizozo. The gate is open between 8:00 A.M. and 2:00 P.M. Visitors receive handouts and are allowed to drive the seventeen miles to Trinity unescorted.

Another way of entering the missile range is to travel with a caravan organized by the Alamogordo Chamber of Commerce. The caravan forms at the Otero County Fairgrounds in Alamogordo and leaves at 8:00 A.M. with a military police escort. The round trip is 170 miles. There are, of course, no service stations on the missile range, so gas-up before you leave Alamogordo.

> Website:  www.wsmr.army.mil/pao/TrinitySite/trinst.htm
> Telephone for White Sands Public Affairs Office:
> 505–678–1134 ext 1700

# White Sands Museum, White Sands Missile Range, New Mexico

At the end of World War II, the leading German rocket scientists, led by Werner von Braun, decided to surrender to the Allies rather than be captured by the Russians. They believed that the United States, not the Soviet Union, had the economic power and political freedom necessary to support a rocket program for space exploration. These German scientists had designed and built the world's first missiles, the V-2 rockets that rained down from the skies over London bringing death and destruction. The "V" stands for "Vergeltungswaffe" (vengeance weapon). Hitler chose the name because he wanted to punish the Allies for bombing German cities. Von Braun was a reluctant warrior. He had intended his rockets for space exploration, but Hitler hijacked them for use in his war machine. When von Braun heard of the first V-2 attack on London, he remarked that the rocket had landed on the wrong planet. Von Braun and his colleagues came to the United States and helped start a rocket program here. This work eventually led to the development of the Saturn-V rocket that sent men to the moon.

The White Sands Missile Range (WSMR), formally known as the White Sands Proving Grounds, was established as a testing site for missiles. In fact, a captured V-2 rocket provided one of the first test launches. Since then, more than 42,000 rockets and missiles have been launched here. NASA regularly launches rockets from WSMR to conduct research on the Sun, stars, and microgravity environments. Today, some of the military's most advanced weapons systems are tested here. For example, the THAAD (Theater High Altitude Area Defense) missile is designed to destroy an incoming missile by actually hitting it rather than blowing up. Scientists are also testing laser-based weapons that destroy enemy aircraft and missiles.

The White Sands Museum includes exhibits on calculating instruments such as slide rules and hand-cranked mechanical calculators, gyroscopes, radiation detection instruments, cameras for recording missile tests, and a model of the Large Blast Thermal Simulator, a facility that simulates the detonation of a nuclear weapon so that the effects on military equipment can be measured. Outside in the "Missile Park," you can see more than fifty rockets, missiles, artillery, and aircraft that have been tested at WSMR; these include the Pershing and Patriot missiles, a Huey helicopter, the Firebee target drone, and a Howitzer cannon. The most unusual item on display has to be the flying saucer-shaped Pepp Aeroshell, used by NASA to test parachutes for the Viking mission to Mars. The most historically significant exhibit is a refurbished V-2 rocket with some side panels removed so that the gyroscope, fuel tanks, and rocket engines can be seen. The V-2 is housed in a building at the side of the park. A free booklet available in the museum has detailed information on the V-2 program at WSMR. If you look between the museum and the missile park toward the desert, you will see the entrance to a short nature trail through the Chihuahuan desert with plaques identifying the flora and fauna.

## Visiting Information

The museum is open Monday through Friday from 8:00 A.M. until 4:00 P.M. and from 10:00 A.M. until 3:00 P.M. on Saturday and Sunday. The missile park is open from dawn till dusk. The museum and park are free. There are vending machines but no food serv-

Website: www.wsmr-history.org
Telephone: 505–678–8824

ice. The WSMR is sometimes closed during missile tests. The WSMR Museum is in south-central New Mexico off Highway 70 about 44 miles west of Alamogordo and 22 miles east of Las Cruces. The exit is well marked.

## Nevada Test Site and the Atomic Testing Museum, Las Vegas, Nevada

The Nevada Test Site (NTS) was established in 1951 as an area to be used for testing nuclear weapons. Covering an area larger than the state of Rhode Island, the NTS was home to 928 nuclear explosions between 1951 and 1992. From 1951 through 1962, about 100 nuclear weapons were tested in the atmosphere. After the Limited Test Ban Treaty prohibiting atmospheric testing was ratified in 1963, the tests were moved underground. The underground tests transformed the desert into a moonlike landscape pockmarked with craters. Perhaps the most unusual project undertaken at the NTS was Operation Plowshare, which experimented with possible peaceful uses of nuclear weapons such as mining, cutting paths through mountains for highways, creating artificial ocean harbors, and digging canals. Operation Plowshare got its name from a Bible verse in the book of Isaiah: "And they will beat their swords into plowshares and their spears into pruning hooks. Nation will not take up sword against nation, nor will they train for war anymore." Unfortunately, the nuclear blasts create too much airborne fallout to be a safe option for most proposed applications.

The underground testing was suspended in 1992 when President George H. W. Bush agreed to a moratorium on underground testing. When the United States agreed to the Comprehensive Test Ban Treaty, the testing came to an end. Today, Los Alamos National Laboratory conducts experiments at the NTS to ensure the viability of the U.S. nuclear stockpile. No test involves a chain reaction, and such tests are allowed by the treaty. On the far western border of the site sits Yucca Mountain where a national repository for nuclear waste is under construction.

The only way to visit the Nevada Test Site is to take the free public tours offered monthly by the U.S. Department of Energy. Check the website for tour dates, registration, and other details. The tours leave from the Atomic Testing Museum at 7:30 A.M. and return at 4:00 P.M. You will travel by a chartered bus equipped with a restroom. There are no stops for lunch so you'll have to bring your own food and drinks. You should wear sturdy shoes, and neither shorts nor sandals are permitted. Participants must be at least fourteen years old. No cameras, binoculars, tape recorders, or computers are permitted. The tours fill quickly so plan as far in advance as possible.

The tour includes the following points of interest: the town of Mercury, which serves as base camp for the test site; Frenchman Flat where the first atmospheric test at NTS took place; the Nonproliferation Test and Evaluation Complex used to test procedures for cleaning up oil and chemical spills; the Low-Level Radioactive Waste Management Site for disposing of radioactive waste from the production of nuclear weapons; Control Point-1, the command post used during nuclear tests; News Nob, the viewing area the press and VIPs used to watch the tests; and Sedan, a cratering experiment that was part of the Plowshare program.

If you want an intellectual break from the cacophonous casinos, sexy shows, and belt-loosening buffets for which Las Vegas is famous, head straight for the Atomic Testing Museum, established to preserve the historic legacy of the NTS. The museum examines the atomic age from World War II to the present with a strong emphasis on the role of the Nevada Test Site. Here, you can see an exhibit on how Las Vegas was affected by the test. Because the mushroom clouds from the atmospheric tests could be seen from Las Vegas, the tests became a tourist attraction, and there was even a Miss Atom Bomb beauty pageant. The museum's Ground Zero Theatre provides a multisensory experience of what a test was like complete with shaking seats and a blast of hot air. In the Underground Testing Gallery, you can see drill bits and other equipment used for the underground tests. There's a display case full of lunch boxes, soft drinks, science kits, and other atomic age memorabilia that reveals how the atomic age influenced popular culture. The museum is replete with video monitors where you can watch vintage footage of nuclear weapons tests. Indeed, most of the famous clips of nuclear explosions that appear on television and in documentary films were filmed at the NTS. At the end of your route through the museum, a chunk of graffiti-decorated concrete from the Berlin Wall symbolizes the end of the cold war while a steel beam from the World Trade Center symbolizes the beginning of the war on terrorism.

### Visiting Information

The museum is open Monday through Saturday from 9:00 A.M. until 5:00 P.M. and on Sunday from 1:00 P.M. until 5:00 P.M. The museum is closed on New Year's Day, Thanksgiving, and Christmas. Admission is $10 for adults 18 and older,

> Websites:  www.atomictestingmuseum.org
> (Atomic Testing Museum)
> www.nv.doe.gov
>
> Telephone:  702–794–5161 (museum)
> 702–295–0944 (site)

$7 for seniors 65 and older, $7 for youth 7 to 17, and children 6 and under are free. The Atomic Testing Museum is located on the first floor of the Desert Research Institute, only about a mile from the fabulous strip. The street address is 755 East Flamingo Road.

## Titan Missile Museum, Tucson, Arizona

From 1963 through 1984, the Titan II missile was the primary ICBM in the U.S. defense arsenal. Standing 110 feet tall and weighing 170 tons when fueled, they were the largest ICBMs ever produced by the United States.

A missile in its silo.

These missiles served as a retaliatory deterrent, to be launched only in response to a Soviet first strike. They were grouped in clusters of eighteen in Kansas, Arkansas, and here in Arizona. Each of the fifty-four missiles carried a nuclear warhead in excess of one megaton and could be launched in less than one minute.

In 1981, President Ronald Reagan approved the Strategic Forces Modernization Plan, which replaced the Titan II missiles with the more advanced Minuteman and MX Peacekeeper ICBMs. The decommissioned Titan rockets were used to send satellites into space. The Air Force designated this missile silo, Site 571–7, for historic preservation.

One-hour guided tours of the missile site begin every half-hour starting at 9:00 with the last tour leaving at 4:00. The tour begins with a video detailing the history of the Titan II. After donning a hard hat, you walk to the top of the launch duct where you can look down at the menacing missile. A two-foot hole was cut in the nose cone, and the door that slid across the top of the silo was permanently sealed halfway open so that Soviet satellites could verify the missile's deactivation. Next, you descend into the hardened launch control center and experience a simulated missile launch. Mounted on giant springs with a foot of space between the inner and outer walls so that everything can move and featuring walkways on top of hydraulic shock absorbers, the launch center is designed to withstand anything short of a direct hit from a nuclear warhead. Here, crews of four—two officers and two enlisted men—worked twenty-four-hour shifts. If the crew received orders to launch the missile, the two officers walked to a locked filing cabinet and retrieved the launch codes. They would then each go to one of two separate control consoles and insert a key. To activate the consoles, the keys had to be turned at the same time. The keyholes were purposely placed too far apart to be operated by a single person. Finally, the guide leads you through two three-ton blast doors to the silo itself for an underground view of the missile. *Star Trek* fans may be interested to know that the missile silo scenes in the movie *Star Trek: First Contact* were filmed here. Back at ground level, the Titan II's rocket engines and nose cone are on display along with some vehicles and refueling equipment.

## Visiting Information

The museum is open from 9:00 A.M. until 5:00 P.M. daily except Thanksgiving and Christmas. Admission is $8.50 for adults, $7.50 for senior citizens, and $5.00 for children ages 7 to 12. Children 6 and

> Website: www.pimaair.org
> Telephone: 520–625–7736

under are free. The museum offers a variety of special tours such as the "Beyond the Blastdoor" tour, where you can see the crew's living quarters and stand at the base of the silo and look up at the missile. Check the website for details and pricing on the special tours. The Titan Missile Museum is located in Sahuarita, Arizona, about twenty miles south of Tucson. From Tucson, take I-19 south and take exit 69.

## Minuteman Missile National Historic Site, South Dakota

During the cold war, this region of southwestern South Dakota was a missile field pockmarked by 150 Minuteman missile silos and 15 launch control centers. Contrary to popular belief, the missile sites were not top secret or disguised in any way. In fact, some sites could be seen from the highway, a few were shown on television programs, and the Air Force even gave tours during "Community Days." What was kept secret were the launching procedures, the targeting, and what specific missiles were armed with warheads. Today, only a single silo and launch center remain, preserved by the National Park Service as a sobering reminder of a time when the possibility of an all-out nuclear war was an everyday reality. The rest of the missiles were deactivated as stipulated in the START Treaty in 1991. A new generation of about 500 Minuteman III missiles is currently deployed in North Dakota, Wyoming, and Montana.

The Minuteman Missiles are named in honor of the minutemen of the Revolutionary War, citizen soldiers who, according to legend, could be ready for battle in less than a minute. The Minuteman II Missiles, in service from 1965 through 1994, could be launched in only five minutes, enabling the United States to respond to a Soviet first strike before the silos could be destroyed. With a maximum speed of 15,000 mph, the Minuteman II could fly over the North Pole and hit a Soviet target 6,000 miles away in about half-an-hour. Each was armed with a 1.2-megaton warhead, the equivalent of 120 Hiroshima bombs. The Minuteman II was the first solid-fueled ICBM and, as a result, it was easier to maintain and much cheaper—about 20 percent of the cost of liquid-fueled predecessors. The missiles were housed in underground reinforced concrete silos sealed at the top by a ninety-ton door. In the event of a launch, explosives would have blown the door off and allowed the missile to rise out of its underground chamber.

The weapons were controlled by "missileers" stationed miles away from the silos at launch control facilities. The life of a missileer has been described as "hours and hours of sheer boredom punctuated by seconds of panic." The launching procedure was similar to that of the Titan missile described in the previous entry. Two keys had to be turned simultaneously by two different crew members. The keyholes were positioned about twelve feet apart so that no single crew member could act on his own accord. Some movies have portrayed the turning of the keys as an angst-filled moment with one or both soldiers, paralyzed with fear, reluctant to turn the key. But the crew members were repeatedly tested to make sure they would follow orders without hesitation, and veteran launch officers assure us that they would have done their duty without reservations.

## Visiting Information

The Minuteman National Historic Site consists of two facilities about fifteen miles apart: the Launch Control Facility, Delta-01, and the actual missile silo, Launch Facility, Delta-09. A glass door covers the top of the silo so visitors can peer down and see the unarmed Minuteman II Missile inside. Self-guided tours of the silo are available by dialing a number on your cell phone. The Launch Control Facility includes an aboveground building with a kitchen, sleeping quarters, and life support equipment connected via elevator to the underground control room. Because the elevator holds just six people, the Launch Control Facility is accessible only on guided tours. These guided tours include both facilities, but reservations must be made in advance—months in advance during the summer. The guided tours start at the Visitor Contact Station where you can view exhibits on the cold war and an orientation film. The station is located off of exit 131

| Websites: | Minuteman Missile NHS: |
|---|---|
| | www.nps.gov/mimi |
| | South Dakota Air and Space Museum: |
| | www.ellsworth.af.mil/museum.asp |
| Telephone: | 605–433–5552 (nhs) |
| | 605–385–5188 (museum) |

on I-90 next to the Conoco Gas Station. The Contact Station is open from 8:00 A.M. until 4:30 P.M. Monday through Saturday during the summer. It is closed on Saturdays during the rest of the year. The Launch Facility is six miles west of Wall, South Dakota, at exit 116 off of I-90. Nearby, the South Dakota Air and Space Museum at Ellsworth Air Force Base has artifacts and displays related to the cold war; you can see a launch control center and silo used for training the crews.

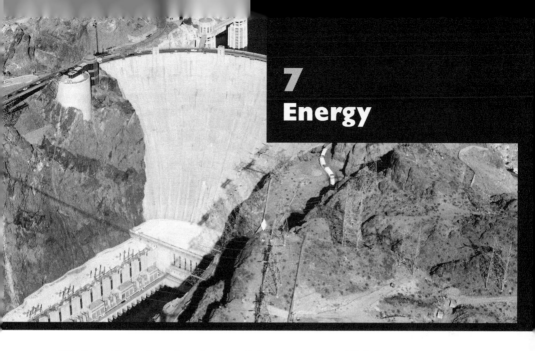

# 7

# Energy

*We do not know how the scientists of the next century will define energy
or in what strange jargon they will discuss it. But no matter what language
the physicists use . . . energy will remain in some sense the lord and
giver of life, a reality transcending our mathematical descriptions.
Its nature lies at the heart of the mystery of our existence as animate
beings in an inanimate universe.*

Freeman J. Dyson

Newspaper headlines warn about an energy crisis, conserving energy is encouraged as a civic virtue, and politicians propose energy plans to lessen our dependence on foreign oil. We hear a lot about energy these days, but what exactly is energy? In physics energy is defined as the ability to do work. Work has a very specific meaning in physics: to *do* work means applying a force on an object and moving that object through some distance. For example, consider a weight that is lifted off the ground and held hovering above a stake. If the weight is released, it can apply a force to the stake and move it through some distance. In other words, the weight can do work on the stake and drive it into the ground. By lifting the weight off the ground, you've given it something it didn't have when it was lying on the ground; you've given it the ability to do work. You've given it energy.

Now you may find this definition of energy unsatisfying because, after all, it tells you what energy can do (work), but the definition doesn't really tell you what energy *is*. The irony is that while energy is one of the most important unifying concepts in all of science, it is impossible to define it in a concrete way. Energy is not like a force that you can feel, or time that you can experience, or a material object that you can touch. Energy is an abstract mathematical quantity that can be calculated. Energy comes in a variety of forms: gravitational, kinetic, elastic, chemical, radiant, nuclear, electrical, thermal, and the energy associated with matter. A mathematical formula calculates each form. Energy is measured in a variety of units including joules, calories, and kilowatt-hours.

The really important thing about energy is that it is conserved. This fundamental law of physics can be stated simply:

> Energy cannot be created or destroyed. Energy can be transformed
> from one form into another. Energy can be transferred from one object
> or place to another. But the total energy in a closed system always
> remains the same.

To fully understand this law, we must first understand what is meant by a "system." A system can be thought of as an imaginary box with a lid containing the matter and energy under study. If the lid to the box is open so that matter and energy can be either put in or taken out of the box, then we have an *open system*. If the lid to the box is closed so that no matter or energy goes in or out, then we have a *closed system*.

As an illustration of the law, suppose a closed system contained the following mix of energies: twenty units of kinetic energy, ten units of gravitational energy, thirty units of chemical energy, and forty units of thermal energy. Add it all up, and there's a total of one hundred units of energy. Now go away for awhile, and let the system do whatever it's going to do. If you come back some time later and examine the system, then the energy may be in a different place (transferred) or a different form (transformed). For example, twenty units of the chemical energy may have burned to produce twenty units of thermal energy, and the ten units of gravity energy may have been transformed into ten units of kinetic energy so now the mix of energies is different: thirty units of kinetic energy, ten units of chemical energy, and sixty units of thermal energy. If you add it all up, there are still one hundred units of total energy. That's the essence of the conservation of energy, one of the most powerful ideas in all of science.

If the conservation of energy is a law of nature, then why are we continually admonished to conserve energy? Energy will be conserved, no matter what we do. The problem involves the forms of energy; some are more concentrated and therefore more useful than others. For example, a gallon of gasoline is a very concentrated form of chemical energy. When the gasoline is burned, the chemical energy is transformed into thermal energy—the wiggling and jiggling motion of atoms and molecules. The atoms and molecules close to the burning gasoline bump into the neighboring atoms and molecules and share some of their energy. Those atoms and molecules, in turn, bump into their neighbors and give up some of their energy and so on. Pretty soon, all that useful, concentrated chemical energy has been dissipated into the surrounding air. All the energy is still there, but it's spread out too thinly to do anything useful. Consider the following analogy: suppose we had $10 to give to charity. Would it do more good to give the entire $10 to a single charity or give one cent to each of 1,000 charities? Clearly, giving one cent to 1,000 charities wouldn't do any of them much good. Strictly speaking, we should be telling people to conserve useful forms of energy, like coal, oil, natural gas, and uranium.

Energy is the lifeblood of modern civilization. When the electrical power goes out, most work comes to a grinding halt. An oft-repeated urban legend claims that exactly nine months after an electrical outage, local hospitals experience a sudden upward blip in the number of babies born (perhaps people do find productive—or, in this case, reproductive—things to do during an outage). Our country is the greatest consumer of energy in the world. In the United States in 2005, 40 percent of the total energy consumed generated electric power, 28 percent powered transportation, 21 percent energized industry, and 11 percent supplied commercial and residential use. Where did this energy come from? About 40 percent came from petroleum (oil), 23 percent from coal, 23 percent from natural gas, 8 percent from nuclear, and 6 percent from renewable sources such as hydroelectricity, wood, geothermal, solar, wind, and waste.

Because the largest chunk of energy goes into generating electricity, let's talk about how that's done. Modern commercial electrical power generation is based on a phenomenon called electromagnetic induction, discovered independently in 1831 by physicists Joseph Henry in the United States and Michael Faraday in England. The scientists demonstrated that electricity can be created from magnetism by simply changing the strength of the magnetic field enclosed by a coil of wire. This can be done by either spinning a coil of wire in a magnetic field or spinning a magnet around a coil of wire. In a commercial power plant, the latter method is preferred. To get the magnet to spin, it is connected by a shaft to a turbine, which consists of a set of

blades arranged in a circle—like the blades of a fan. The blades of the turbine are rotated by steam (or, in the case of hydroelectricity, flowing water).
Steam is created by boiling water. Boiling water requires a source of heat. To
get heat, something—coal, oil, natural gas, or uranium—must be burned. To
summarize: something is burned to make heat to boil water to make steam.
The steam turns the turbine that rotates a magnet around a coil of wire
and—violá—electricity! The energy used for electrical power generation
breaks down as follows: 49 percent from coal, 20 percent from natural gas,
19 percent from nuclear, 7 percent from hydro, 3 percent from renewables,
and 2 percent from fuel oil.

A quick glance at the statistics cited for total energy reveals two major
problems afflicting our country and the larger world. First, a huge chunk of
our energy comes from oil, much of which is imported from unstable and
often unfriendly countries in the Middle East and elsewhere. We are too
dependent on these countries for our energy requirements, and we need to
find ways to become energy independent. Because most petroleum is used
for gasoline for our cars, the solution to this problem involves producing
more fuel-efficient cars, finding alternative fuels for them to run on, or running them on electricity.

The second problem is that fully 86 percent of our energy comes from
burning coal, oil, and natural gas. When these fossil fuels are burned, they
produce greenhouse gases, such as carbon dioxide ($CO_2$). These gases can
increase the earth's temperature in a process called global warming. The
mechanism behind global warming works like this: the earth's atmosphere
is largely transparent to visible wavelengths of light coming in from the sun.
When the earth absorbs this energy, it re-radiates much of it at wavelengths
in the infrared part of the spectrum. Gases such as water vapor and carbon
dioxide are good absorbers of infrared radiation. These gases absorb the
infrared and then re-emit it in all directions including back down toward the
surface of the earth. Thus, the energy gets trapped and the temperature goes
up. A similar process happens when you park your car in the summer sun.
The glass allows the visible sunlight through to the interior of your car. The
car interior re-emits the energy at infrared wavelengths which the glass
absorbs. Again, the energy gets trapped, and the interior temperature rises.

There has been much political controversy regarding global warming,
yet it is indisputable that during the last century the average temperature at
the surface of planet Earth has increased by 1°F. The last century has also
seen the sea level rise by six to eight inches, partly because of melting gla-

cial ice and partly due to the expansion of sea water. There now appears to be a clear scientific consensus: most of this warming has resulted from the increase in greenhouse gases produced through human activities. Less certain is by how much and how fast the temperature will continue to rise. The Intergovernmental Panel on Climate Change is projecting anywhere from a 2.2 to 10°F increase by 2100.

Below, I describe each major source of energy along with some places you can go to learn more about them. The sources are listed in order of current importance, beginning with the largest source, oil. After that, I describe a couple of specific sites, Three Mile Island and the National Renewable Energy Lab.

## Oil

Five thousand years ago, early Middle Eastern civilizations used asphalt formed from surface oil for mortar and waterproofing. The ancient Egyptians used liquid oil for medicinal purposes and embalming. Archaeological evidence suggests that the Chinese were using bamboo poles to dig 800-foot-deep oil wells as early as 347 A.D. A new use for oil was discovered in 1849 when Canadian geologist Abraham Gesner distilled a lamp fuel he called kerosene from petroleum. Demand for the cleaner and cheaper kerosene increased while the demand for whale oil decreased, perhaps saving the whales from extinction. Although Gesner never reaped any financial reward for his discovery, he later became known as the "Father of the Petroleum Industry."

The first oil well in the United States was drilled by "Colonel" Edwin L. Drake near Titusville, Pennsylvania, an event that marked the birth of the oil industry. Almost overnight, the population of Titusville grew from 250 to 10,000, and the owner of the land became the first oil millionaire. The industry suffered a setback in the late 1800s when Thomas Edison's electric light bulb began replacing kerosene lamps. But the industry rebounded in the early 1900s when Henry Ford started mass-producing his Model T automobile, fueled by a previously little-used petroleum by-product called gasoline. At about the same time, the Spindletop oil field near Beaumont, Texas, produced a "gusher" that shot oil 160 feet into the air. It spewed for nine days until the contractor figured out how to cap the well. Once under control, the well was producing an unheard of 100,000 barrels of oil every day. This event marked the beginning of the Texas oil boom. Within months,

An oil derrick pumping oil out of the ground.

40,000 boomers poured into Beaumont, and within a year 280 oil wells dotted the pumping site. Later in the twentieth century, chemists invented synthetic materials such as plastics that further increased the demand for oil.

Today, the worldwide demand for petroleum has reached about 80 million barrels of crude oil per day. The United States, with roughly 5 percent of the world's population uses about 25 percent of the global supply. This means that every man, woman, and child in the United States consumes an average of three gallons of crude oil every day. Because of increases in oil prices, worries about U.S. dependence on foreign oil, and concern for the natural environment, a movement is afoot to wean us away from oil.

### Sites and Visiting Information

You can go back to the very beginning of the U.S. oil industry at the Drake Well Museum in Titusville, Pennsylvania. The main attraction here is an exact working replica of Drake's oil derrick and pump house, a design that became the industry standard around the world. The museum houses more than two thousand oil-related artifacts, including machinery, tools, engines, and drilling rigs. Outside, a nitroglycerin wagon that carried explosives used to break up oil-bearing rock is on display along with a photographic wagon and its own dark room. Oil-collecting pits used by prehistoric people can also be seen. The museum is open Monday through Saturday from 9:00 A.M. until 5:00 P.M. and on Sunday from noon until 5:00 P.M. It is closed on Mondays from November through April. Admission is $5 for adults, $4 for seniors, and $2 for youth.

The Spindletop-Gladys City Boomtown Museum, just outside of Beaumont, Texas, is unique among the many places you can visit to learn about oil. Operated by Lamar University, this living history museum is a recreation of the town that sprang up after the gusher commenced its gushing. The fifteen buildings—all furnished with period furniture and artifacts—include a drug store, general store, post office, livery stable, barbershop, and, of course, a saloon, where legend has it that at least one man was killed every Saturday night. A reproduction of a sixty-four-foot-tall oil derrick stands at the center of the town. The museum is open Tuesday through Saturday from 10:00 A.M. until 5:00 P.M. and on Sunday from 1:00 P.M. until 5:00 P.M. Admission is $3 for adults and $1 for children. A visit here could easily be coupled with a trip to the Texas Energy Museum in Beaumont, where you can learn more about petroleum geology, energy, and science.

The Petroleum Museum in Midland is home to the forty-acre "oil patch" collection, the world's largest assemblage of historic drilling equipment and modern petroleum-related machinery. The museum has three wings dedicated to geological, technical, and cultural exhibits that explore every facet of the oil industry. Life in the oil fields is related through photographic murals and taped interviews with some petroleum pioneers.

Several major science museums have permanent displays about petroleum, most notably the "Petroleum Planet" exhibit at Chicago's Museum of Science and Industry and the Weiss Energy Hall in the Houston Museum of Natural Science. For a complete listing and brief descriptions of sites in the United States relating to the oil industry, visit the website of the American Oil and Gas Historical Society.

> **Websites:** Drake Well Museum:
>    www.drakewell.org
> Spindletop Museum:
>    http://spindletop.org
> Texas Energy Museum:
>    www.texasenergymuseum.org
> American Oil and Gas Historical Society:
>    www.aoghs.org
>
> **Telephone:** Drake Well Museum:   814–827–2797
> Spindletop Museum:   409–835–0823
> Texas Energy Museum: 409–833–5100

# Coal

Humans have used coal ever since an observant cave man noticed that a certain type of black rock would burn. Archaeological evidence indicates that Roman soldiers burned coal in their campfires in England during the second and third centuries. In the 1300s, the Hopi Indians used coal for heating, cooking, and baking pottery. Colonial blacksmiths used coal imported from England and Nova Scotia to fire their furnaces. The first clear, historical record of coal in the territories shows up on a 1637 map of the Illinois River drawn by French explorers. In 1748, commercial coal production in the colonies got its start in mines near Richmond, Virginia. During the Industrial Revolution, the use of coal increased exponentially as steam-powered machines replaced human and animal muscle. By 1875, coke, a by-product of coal, was replacing wood charcoal as the preferred fuel in iron blast furnaces. The use of coal to generate electricity can be traced to 1882, when Thomas Edison built the world's first coal-fired electrical generating station in New York City. Today, about half of the electrical energy generated in the United States comes from burning coal.

## Sites and Visiting Information

Producing coal from 1855 to 1972, the No. 9 Coal Mine is advertised as the world's oldest continuously operated anthracite coal mine. At its peak, 450 men representing a melting pot of ethnic groups slaved away deep in the mine. The museum holds a jumble of their equipment ranging from drills, picks, and shovels to detonators and blasting caps. A coal-fired furnace, a "mucking machine," and a replica of a miner's kitchen are also on display. Wire baskets suspended from the ceiling hold mining clothes just as the "wash shanty" did when the mine was operating. Photographs of dirty, coal-dust-covered miners hint at the conditions they endured. If time allows, you can view an hour-long documentary about coal. The highlight of your visit is a ride into the mine on a coal train. The train takes you 1,600 feet into the side of a mountain where you disembark and are led on a walking tour. The guide's stories tell what it was like to work in the mine. The route takes you past a 900-foot elevator shaft, a "muleway," and a miner's hospital cut into the rock.

The #9 Coal Mine is located in Lansford, Pennsylvania, just off of Route 209 at 9 Dock Street. The museum is open all year, but the mine tours run from May through October. Mine tours are given Friday, Saturday, and Sun-

Sue Smith/Shutterstock

Carts used to haul coal out of the mines.

day from 10:00 A.M. until 4:00 P.M. During June, July, and August, Thursday tours are available. After Labor Day, the tours are limited to weekends. Tours begin approximately on the hour with the first tour leaving at 11:00 A.M. and the last tour at about 3:15 P.M. Admission to the mine and museum is $7. The temperature inside the mine is about 54°F so bring a sweater or jacket. (If you forget, the museum has jackets you can borrow.)

Other coal mine tours operating in Pennsylvania include the Pioneer Coal Mine in Ashland and

> Website:  http://n09mine.tripod.com
> Telephone:  570–645–7074

the Lackawanna Mine in Scranton. Several other states—West Virginia, Kentucky, Illinois, and Wyoming—have coal mine tours.

## Nuclear Energy

The 104 nuclear reactors at sixty-five sites across thirty-one states currently provide nearly 20 percent of the electricity in the United States. The major difference between a nuclear power plant and any other steam-driven electric generating plant be it coal, oil, natural gas, wood, or the sun, is the source of heat used to boil water. In a nuclear power plant, that heat comes from nuclear fission reactions where tiny amounts of matter are converted into huge amounts of energy ($E=mc^2$). The heart of a nuclear power plant is its nuclear reactor, which consists of four main components: fuel rods, control rods, the moderator, and the coolant.

In the fuel, uranium, the percentage of the fissionable isotope U-235 has been increased to 3 percent. Because the U-235 fuel is so highly diluted, it is impossible for a commercial nuclear reactor to explode like a nuclear bomb. The uranium is formed into pellets about the size of the last joint in your pinky finger and stacked one on top of another inside long thin metal rods. Tens of thousands of these fuel rods containing about one hundred tons of fuel form the heat-producing reactor core. Each pound of uranium fuel holds nearly the same amount of energy as fourteen freight car loads of coal. Thus, a nuclear reactor need be refueled only once every year or two.

The fission is produced by neutrons that are released when the nuclei split. So, by controlling the number of neutrons, the chain reaction can be controlled. This task is accomplished with control rods made of materials, usually cadmium or boron, that absorb neutrons well. The control rods can be raised or lowered into the reactor to speed up, slow down, or stop the reaction. The fuel rods and control rods sit inside a steel reactor vessel engineered to withstand high pressures, to absorb neutrons, and to shield against radiation.

The neutrons are more likely to cause fission if they are slowed down after they are expelled from the nuclei by having the neutrons collide with particles of comparable mass. Protons have about the same mass as neutrons, and single protons form the nucleus of hydrogen. What cheap material has hydrogen in it? Water! The speed of the neutrons is slowed or "moderated" by hitting the hydrogen nuclei in the water molecules.

The final component is the coolant that circulates through the reactor, absorbs the heat, and carries it away to boil water. What's a fluid material that can hold a lot of heat? Once again, water! In most reactors, this coolant water is kept under high pressure so it can attain high temperatures without boiling. There are two separate loops of water in a nuclear power plant. The coolant water forms the primary loop, which transfers the heat to a secondary loop, causing it to boil and change to steam. This heat exchange is accomplished by placing the pipes carrying the coolant water in contact with the pipes carrying the water in the secondary loop. Once the coolant water releases its heat it is pumped back into the reactor.

The reactor core, the vessel, and the coolant water loop become radioactive during the operation of the reactor and are therefore enclosed in con-

A nuclear power plant. The domed buildings hold the nuclear reactors.

crete. The entire apparatus sits inside an airtight containment building made from steel reinforced concrete where the walls measure three feet thick. In fact, the containment buildings at nuclear power plants are among the strongest structures ever made and are designed to withstand an impact from a commercial jet. Because of all this shielding, the radiation exposure outside a nuclear plant is minuscule, barely above the background radiation from natural sources.

The worst type of accident that could happen at a nuclear power plant is a loss of the coolant water to the core. But remember that the coolant water serves another purpose: it acts as a moderator. If a loss of coolant water occurs, then the moderator is also lost. The neutrons move too fast to cause fission, and the reaction shuts down automatically. The bad news is that the fuel is still hot from the radioactivity and if it gets too hot, it can melt. This event is called a meltdown and the accident at Three Mile Island is an example.

Nuclear power has several advantages over other ways of generating electricity: no air pollution is produced, no greenhouse gases are emitted, and therefore there is no contribution to global warming; the United States has enough uranium to fuel our reactors for 200 years, assuming the current rate of consumption; new reactor designs eliminate any possibility of a meltdown; and finally, the energy is in a very concentrated form so the environmental impact of mining uranium is much less than that of coal. Critics point to the high cost of building a nuclear power plant, but these costs can be dramatically reduced by streamlining the licensing procedure and by agreeing on a standardized design. The nuclear power industry has an excellent safety record. In fact, in the entire history of commercial nuclear power in the United States, no one has died from exposure to radiation—no one!

Currently, the main objection to nuclear power centers on the problem of what to do with the high-level radioactive waste that the industry produces. The waste is stored at the reactor site for several decades to allow the short-lived radioactive isotopes to decay. But after that, the long-lived isotopes must be isolated from the environment for thousands of years. Current plans call for the waste to be incorporated into stable glass, encasing the glass in several layers of steel and concrete, and burying the material in stable rock formations. A permanent site for the disposal of this waste is under development at Yucca Mountain in Nevada. Nuclear waste is a political, rather than a scientific, problem. Nobody wants the stuff buried anywhere near them (as the acronym NIMBY proclaims: Not In My Back Yard!). The volume of nuclear waste could be drastically reduced by recycling the spent fuels, extracting the fissionable materials, and placing them back into the reactor. The French, who get 80 percent of their electricity from nuclear power, recycle their fuel as do several other countries. Recycling nuclear fuel is currently against the law in the United States because of concerns about nuclear proliferation.

### Sites and Visiting Information

After 9/11 many nuclear plants overreacted by closing visitor centers and eliminating public tours. For an industry trying to improve its public image, these actions are counterproductive. In France, tours of nuclear facilities provide a popular vacation activity. Nevertheless, a few plants retained their touring programs, but these tours have restrictions, must be arranged well in advance of a visit, and may require a minimum number of people. Among the visitor-friendly sites are the Palisades Power Plant in Covert, Michigan, and the San Onofre Nuclear Generating Station north of San Diego. The Seabrook Plant in New Hampshire has a visitor center with displays on nuclear energy. For a complete listing of all the nuclear plants in the United States, see the website below.

The first city to be lit up, albeit temporarily, by nuclear power was Arco, Idaho, in 1955. The reactor that supplied the power was the Experimental Breeder Reactor #1 (EBR1), now part of the Idaho National Engineering Laboratory, sixteen miles south of town. Although most of the lab is closed to the public, tours of the EBR1 are available. Tour request forms must be completed at least a month in advance. The Museum of Idaho in Idaho Falls has an exhibit called "Race for Atomic Power" that traces the history of the development of nuclear power with an emphasis on the contributions made in southeastern Idaho. The museum is open on Mondays and Tuesdays from 9:00 A.M. until 8:00 P.M. and from 9:00 A.M. until 5:00 P.M. the rest of the week. The museum is closed on Sundays. Admission is $6 for adults.

> **Websites:** A listing of nuclear plants
> http://www.eia.doe.gov/cneaf/nuclear/
> page/at_a_glance/states/statesil.html
>
> **Tours of EBR1**
> https://inlportal.inl.gov/portal/server.
> pt?open=514&objID=255& mode=2

Several other sites described elsewhere in this book are significant in the history of nuclear energy. Of particular importance are Argonne National Laboratory in chapter 4 and Oak Ridge National Laboratory in chapter 7.

## Hydroelectricity

Hydroelectricity involves damming up a river and letting the water that falls through the dam turn the turbines. In physics terminology, the gravitational potential energy of the water sitting above the dam is converted into the kinetic energy of the turbines. This produces 7 percent of the electrical energy used in the United States. The major difference between hydroelectricity and a coal-fired power plant or a nuclear power plant is that nothing is burned to boil the water to change it to steam; the water remains in a liquid state. Thus, hydroelectricity is a squeaky clean source of energy that creates no air pollution, toxic waste, or chemical runoff. So why not use hydroelectricity exclusively? Because there are limited numbers of both rivers and places those rivers can be dammed. In the United States, we have already reached the capacity for hydroelectricity.

### Sites and Visiting Information

The most famous U.S. dam is the Hoover Dam situated on the Nevada/ Arizona border. Designated as one of America's seven modern engineering

An aerial view of Hoover Dam.

wonders, Hoover Dam stands 726 feet above the Colorado River making it the second tallest dam in the United States. (The tallest is the 770-feet tall Oroville Dam on the Feather River in California; the tallest dam in the world is the Rogun Dam on the Vakhsh River in Tajikistan at 1099 feet tall.) Built between 1931 and 1936 during the Great Depression, the dam is named in honor of Herbert Hoover, who supported the project first as the secretary of commerce and then as president. The dam was built mainly to control flooding caused by snow melting in the Rocky Mountains and draining into the river. It also dependably provided water for heavily populated southern California and for farm irrigation. Generating electricity was an added benefit.

A total of 21,000 men worked on the dam with 3,500 workers at the site on an average day. The "hard hat," now used by construction workers everywhere, was invented for and first used by these workers. Nevertheless, ninety-six men died from construction-related accidents (falling, drowning, blasting, rock slides, equipment accidents, and so on) during the project. Contrary to popular belief, no bodies are buried in the concrete. A total of 4,360,000 cubic yards of concrete were used to form the dam and the surrounding structures—enough to pave a two lane highway from New York to San Francisco. Hoover Dam's seventeen electrical generators have a generating capacity of about 2000 megawatts, which provides enough energy for 1.3 million people. The cost of the dam was $49 million, which translates into $676 million today. Part of the dam's allure lies in its elegant Art Deco style featuring sculptured turrets rising from the face of the massive structure.

Hoover Dam is a major tourist attraction, drawing around one million visitors each year. As a result, U.S. Highway 93 that approaches the dam is often clogged with traffic, especially on weekends and holidays. The parking garage opens at 8:00 A.M. and charges $7 for parking (cash only). The Visitor Center is open from 9:00 A.M. to 6:00 P.M. daily except Thanksgiving and Christmas. To avoid crowds, arrive early or late. Pedestrians are not allowed on the top of the dam after dark.

There are three touring options for a visit to Hoover Dam. First, a self-guided tour of the Visitor Center includes a ten-minute film in the theater,

Inside the casings of generators at Hoover Dam, electromagnets spin around coils of wire to generate electricity.

the exhibit gallery, the observation deck, and the street-level displays. Admission to the Visitor Center alone is $8. The "Powerplant Tour" includes the Visitor Center attractions plus a thirty-minute guided tour of the penstocks (the giant pipes that deliver water to the generators) and a view of eight of the generators. Tickets for this option cost $11 for adults and $9 for seniors over 62 and juniors ages 4 to 16. Finally, the "Hoover Dam Tour" includes everything in the Powerplant Tour plus a walk through the inspection galleries, a maze of tunnels used to inspect the dam. This tour is quite a bit more expensive at $30. The Dam Tour is not wheelchair accessible. The tours are restructured fairly often so be sure to check the website for up-to-date information. Extending behind the dam for more than one hundred miles is Lake Mead, the largest artificial lake in the United States. Looking at the lake, you notice a white layer above the shoreline resembling a bathtub ring. This is a mineral deposit left by high water caused by unusually high precipitation that hit the western states in 1983.

Nearby, Valley of Fire State Park, on a northern arm of Lake Mead, is known for its bizarre and colorful rock formations. Several other large dams in the United States. These include

> Website:  www.usbr.gov/lc/hooverdam
> Telephone:  866–730–9097

the Grand Coulee Dam in eastern Washington (this is the largest U.S. hydroelectric plant with a generating capacity of 6,809 MW), the Glen Canyon Dam in Arizona, and the Shasta Dam in northern California. Many other dams feature visitor centers.

## Solar Energy

Virtually all of the Earth's energy can be traced back to the Sun. Solar energy is inexhaustible; as long as the Sun is shining, we'll have solar energy. Solar energy is the largest available energy source, and yet it currently accounts for only a tiny fraction of the electricity generated in the United States. Why? The main problem is that sunlight is spread over the entire daylight surface of the Earth. Thus, to produce electricity on a large scale, this diffuse energy must be collected over a very large area, and the sites chosen to build a solar power station must get lots of sunlight. Solar energy makes sense in the desert, but not in Seattle. Solar power's main advantage is that it is environmentally benign, producing no pollution and no greenhouse gases. Also, because solar power plants use readily available materials and modular

MaxFX/Shutterstock

In this trough design, parabolic mirrors focus sunlight onto a pipe carrying oil.

equipment, they are simpler, quicker, and easier to build than a nuclear or coal plant.

Electricity is produced from sunlight by converting it into thermal energy and then using that thermal energy to boil water to make steam that drives a turbine. Today's commercial solar power plants utilize two different designs. The predominant system uses long troughs of mirrors with a parabolic cross-section. The mirrors concentrate the sunlight along the focal line of the mirror where a pipe containing synthetic oil is positioned. The sunlight heats the oil to temperatures greater than 700°F. The hot oil is then pumped to a heat exchanger where the oil transfers its energy to water, which causes it to boil. A second, less common and less efficient system uses sun-tracking mirrors to reflect light up onto a tank of fluid sitting atop a tall tower in the center of the array of mirrors. Of course, the mirrors get dirty so they have to be cleaned periodically by automatic methods; sometimes the mirrors break, often because of the wind, and must be replaced.

### Visiting Information

The best place to see solar power at work is in California's Mojave Desert in and around Barstow. With an average of 340 days of sunshine annually, this region gets nearly twice as much sunshine as most of the rest of the country. Clustered here are nine solar power plants, constructed in the 1980s, known collectively as the Solar Energy Generating Systems (SEGS). Built by Luz Industries, SEGS is operated and partially owned by Florida Power and Light. SEGS I–II with a generating capacity of 44 Megawatts are located at Daggett, SEGS III–VII with a capacity of 150 Megawatts are at Kramer Junction, and SEGS VIII–IX with a capacity of 160 megawatts are at Harper Lake. With nearly a million mirrors harvesting sunlight, SEGS has a combined generating capacity of 354 megawatts, making it the largest solar power installation in the world. In fact, even by themselves, the Harper Lake facilities are the largest solar plants in the world.

Another solar technology that produces electricity is photovoltaics. Photovoltaic cells use the photoelectric effect to transform sunlight directly into electricity, skipping all the business with water, steam, and turbines. The photoelectric effect happens when light shines on a surface, transferring its energy to electrons and causing them to move. In photovoltaic cells, the surface is a semiconducting material, and the moving electrons cause the cell to behave like a battery. You may have a calculator that is powered by photovoltaic cells (the cells are the dark strip above the buttons). Photovoltaic cells are becoming commonplace; they power street signs and entire buildings. Photovoltaics are particularly useful in remote locations where plugging into the power grid is impossible. Large-scale photovoltaics in power plants are also possible. Currently, the largest photovoltaic system in North America is at the Nellis Air Force Base in Clark County, Nevada. There, 70,000 solar panels produce up to fifteen megawatts of power. Florida Power and Light is constructing the world's largest photovoltaic facility, with a planned capacity of twenty-five megawatts, at a site east of Sarasota, Florida.

In addition to driving by the sites described above, the American Solar Energy Society sponsors occasional tours of solar facilities throughout the United States. Also, the

Websites:  American Solar Energy Society:
www.ases.org
Florida Solar Energy Center:
www.fsec.ucf.edu

Telephone:  Florida Solar Energy Center Tours:
321–638–1015

Florida Solar Energy Center offers free public tours on the second Thursday of the month at 2:00 P.M. The building claims to be one of the most energy efficient commercial buildings in the world, and the tour guides point out its features.

# Wind

Wind is caused by the uneven heating of the Earth's surface by the Sun. For example, equatorial regions are heated more than polar regions, and land heats up (and cools down) faster than water. Air over the warmer areas rises, while air over the cooler areas falls, creating a circulation of air known as a convection cell. In this way, somewhere between 1 percent and 3 percent of the sunlight that hits the Earth is transformed into wind energy.

For hundreds of years, humans have harnessed the power of the wind to turn windmills. The first practical windmills, built in Afghanistan during the seventh century, were used to grind corn and draw water from wells. Beginning in the fourteenth century, the Dutch used windmills to drain the Rhine River delta. On the American frontier, windmills were used to crush grain, pump water, and cut wood. In Scotland James Blyth in 1887 built the first windmill to generate electricity and charge batteries. The next year, Charles Brush built the first windmill to produce electricity in the United States in Cleveland, Ohio. By the 1930s, electricity-producing windmills were commonplace on farms and in rural areas unreachable by the electrical distribution system. As the electrical grid expanded into every corner of the country, these windmills disappeared.

Wind can be used to generate electricity by allowing the moving air to rotate the blades of a turbine. The blades face into the wind at an angle. On one side of the circle of rotation, the blades are angled up. The blades force the air downward, and, by Newton's third law (for every action there is an equal and opposite reaction), the air forces the blade upward. When the blades rotate around to the other side, the blades are angled down. The blades force the air upward, and the air forces the blades downward. The result is a continuous rotation. The turbine powers a generator that produces electricity. Modern wind turbines are gigantic, standing twenty stories tall with a trio of blades as long as a football field. To generate large amounts of electrical power, hundreds of wind turbines are spread across thousands of acres forming a "wind farm." These wind farms are sited in open areas unobstructed by trees where the average wind speed is greater than 10 mph.

Although some people object to the aesthetics of having huge wind turbines dotting the countryside, wind power is a renewable source of clean energy. No pollutants or greenhouse gases are produced, and the cost is competitive with other energy sources. For these and other reasons, wind power is the fastest growing form of electrical power production in the United

A modern wind turbine standing in stark contrast to an old farm windmill.

States. If this trend continues, we will soon surpass Europe as the world's leader in wind energy. In 2008, wind power generated 48 billion kilowatt-hours of electricity, about 1.5 percent of the U.S. total.

## Visiting Information

The largest wind farms are found in Texas, California, and the Great Plains. As of 2008, the largest wind farm in terms of generating capacity was the Horse Hollow Wind Energy Center in Taylor and Nolan counties in Texas, about fifteen miles southwest of Abilene. This farm has 421 wind turbines that generate 736 MW of power, enough to supply 220,000 homes. The Fowler Ridge Wind Farm in Benton County, Indiana, with a planned capacity of 750 MW, is currently under construction. Even larger wind farms are on the drawing board.

With more than 4,000 wind turbines scattered across the hilltops, California's Altamont Pass is home to the largest concentration of wind-powered turbines in the world with a generating capacity of a billion kilowatt-hours of electrical energy per year. Most wind turbines you can see here were installed in the early 1980s as one of the first wind farms built in response to tax incentives adopted for alternative sources after the energy crisis of the late 1970s. From the highway, you can see several kinds of wind turbine, including some mounted vertically to resemble giant egg-beaters. Larger and more cost-effective modern wind turbines are gradually replacing old and obsolete turbines here. These small turbines rotate at a relatively high speed and have been criticized by animal activists because the blades can hit and kill birds. The winds that blow through Altamont Pass are generated when the hot air and low pressure in the California Central Valley sucks in cooler air from the ocean. The pass is about a one-hour drive east of San Francisco on I-580 between Livermore and Tracy.

There are no visitor centers at any of these wind farms. They can be viewed while driving along the highway.

## National Renewable Energy Laboratory (NREL), Golden, Colorado

The NREL is the U.S. Department of Energy's main facility for researching renewable energy sources and energy efficiency. So what is a "renewable" as opposed to a "nonrenewable" energy source? A renewable energy source is either continuously available, like sunlight, or replaceable within a human lifespan, like wood. The processes that produce oil, however, take millions of years, well beyond a human lifetime, so fossil fuels along with uranium for nuclear power, are considered nonrenewable. Renewable energy sources include solar, wind, biomass (wood, etc.), hydroelectric, and geothermal. Major research and development areas at the NREL include photovoltaics, bioenergy, and wind technology. Scientists also conduct research on how to make buildings and vehicles more energy efficient.

### Visiting Information

The NREL has a Visitors Center with interactive exhibits on solar, wind, biomass, and other renewable energy sources. There are no public tours of the actual research facilities. Several outdoor exhibits demonstrate solar power and energy saving strategies for your home. Perhaps the most interesting exhibit to see here is the building itself, which is a model of energy-efficient design. The most striking architectural feature is the tall V-shaped glass wall which lights and heats the main exhibit hall. On the opposite side of the windows is a thick concrete wall painted black. The black wall has a layer of glass in front of it separated from the wall by a layer of air. The wall absorbs the sunlight, re-radiates it at infrared wavelengths and warms the air. Other energy-efficient features of the building include the exterior insulation, evaporative cooling, and a computer-controlled energy

> Website:   www.nrel.gov
> Telephone:   (303) 384–6565

management system. The electricity for the building is generated by wind turbines in northern Colorado. The Visitors Center is open Monday through Friday from 9:00 A.M. until 5:00 P.M.

## Three Mile Island Nuclear Power Plant, near Harrisburg, Pennsylvania

The most serious accident in the history of the commercial nuclear power industry in the United States happened at the Three Mile Island Unit 2 plant on March 28, 1979. At around 4:00 A.M. that morning, the pumps in the water loop that carried steam to the turbines failed. By design, this failure automatically triggered the control rods to be lowered into the reactor, halting the fission reaction, and thereby shutting down the reactor. However, stopping the fission doesn't eliminate all the heating because about 5 percent of the heat comes from radioactivity. This radioactivity comes from the decay of radioactive nuclei that are produced during the fission process.

Thus, even though the reactor shut down, water must continue to flow through the loops to prevent overheating.

A backup pump took over to keep the water circulating, but a valve in the pump had been mistakenly left closed. A warning light that would have alerted operators about the valve went unnoticed because it was hidden behind a tag. This caused an eight-minute delay during which the temperature and pressure in the primary loop continued to increase. This resulted in the opening of a pressure relief valve in the primary loop. This valve stuck in the open position, and the malfunction went unnoticed for two hours. This open valve allowed radioactive water to spill out from the reactor, flooding the floor of the containment building. A portion of this water was automatically pumped to another building releasing a small amount of radiation in the process.

The water remaining in the hot reactor began to evaporate. To prevent overheating due to a loss of water, all reactors have a tank of emergency cooling water ready for use in case of an accident. The decrease in the water level in the reactor automatically triggered the release of this emergency cooling water. Unfortunately, the operators misinterpreted their control room instrumentation and concluded that there was too much water in the reactor, rather than too little. In an act that sealed the fate of the doomed reactor, they shut off the emergency cooling water. The water level dropped below the top of the fuel rods, and the fuel began to melt. Because the melting causes the fuel rods to slump downward, this event is called a meltdown. It was later discovered that about half of the core melted. The water that remained at the bottom of the core prevented a complete meltdown in which molten fuel would have spilled onto the floor of the containment building.

By the end of the day, things appeared to be under control, and the reactor seemed to be stable. But by the next morning, new concerns arose regarding how much radiation had been released by the water that had been pumped out of the reactor. Because of uncertainties about the level of radiation, the governor of Pennsylvania announced that he was advising pregnant women and young children, those most vulnerable to the effects of radiation exposure, to evacuate the area within a five-mile radius of the plant.

Later the same day, it was discovered that unusual chemical reactions in the reactor had created a large bubble of hydrogen gas at the top of the reactor vessel. If the hydrogen exploded, it might rip open the reactor vessel, dump radioactive material into the containment building, and possibly create a breach of containment. The crisis finally ended on the afternoon of April 1, when experts agreed that the hydrogen bubble could not explode because of a lack of oxygen in the vessel. By that time, the size of the bubble had been drastically reduced. Earlier that day, after assurances that even if the bubble did explode, it wouldn't do so for a day or two, President and Mrs. Jimmy Carter toured the facility in an effort to reassure the public.

In spite of the serious nature of the accident, there were no immediate deaths. Estimates put the average radiation dose to the two million people in the area at about one millirem. For comparison, a chest x-ray gives an exposure of about six millirem, and the average annual exposure for a U.S. citizen is about 350 millirem. This very small additional exposure might result in one additional long-term cancer death among the population near the plant. The accident cost the utility company $1 billion for the clean up, and they lost an expensive reactor. The accident had a devastating effect on the U.S. nuclear power industry, a setback from which it is only now beginning to recover.

## Visiting Information

The Three Mile Island nuclear power plant is located about ten miles south of Harrisburg, Pennsylvania. See the Pennsylvania Highways website for directions. The first place to stop along PA 441 is the main gate for Unit 1. If you are coming from the north, the sign will be on your right. There is a small visitor parking area to the right. You are not allowed to drive across the bridge. From here, you can see three of the four cooling towers. The function of the cooling towers is to allow heat to be released from the water that circulates through the condenser so that it can be recirculated or flushed into the river. Notice that the two cooling towers closest to you are belching steam. This is because Unit 1 remains an operating nuclear reactor. It was restarted in 1985, six years after the accident. The farthest cooling tower is not releasing steam because it is part of Unit 2, where the accident happened. Get back in your car and continue south on PA 441.

When you see a building on your left with a blue awning, pull into the parking lot. At this training center the utility company, AmerGen, trains its nuclear plant operators. (Both units were originally owned by GPU/MetEd. Now, AmerGen owns Unit 1.) Here, you can see a yellow emergency siren similar to the seventy-nine sirens sprinkled around a ten-mile radius of the plant. In the front of the building are boxes containing continuous air samplers and dosimeters that keep tabs on the air quality and the radiation levels. Out near the road is a Pennsylvania Historical Marker that summarizes the accident. Across the road you see the two inactive cooling towers and to the right of the towers, two concrete

> Website: www.pahighways.com/features/threemileisland.html

structures with a domed roof. These are the containment buildings for the nuclear reactors. The Unit 2 reactor is the one on your left.

## Yucca Mountain, Nevada

As I mentioned in the entry on nuclear energy, Yucca Mountain is the proposed burial ground for high-level radioactive waste from nuclear power plants and national defense facilities. Currently, this waste is being stored at more than a hundred sites scattered across the country, but these facilities are running out of room. The purpose of the Yucca Mountain repository is to provide a permanent site where the waste can be isolated from the environment for at least 10,000 years—enough time for the radioactivity to die down to a safe level. The plan is to encase the waste in multilayer stainless steel and nickel alloy packages protected by titanium shields and bury it in tunnels 1000 feet below the surface of the mountain and 1000 feet above the water table.

Yucca Mountain was chosen for the site because it is geologically stable; it is unlikely to be affected by earthquakes, volcanoes, or other geological events. Also, the mountain (actually a ridge-line) is composed of a volcanic rock called "tuff" with certain chemical, physical, and thermal properties that make it a suitable material for burying radioactive waste in. Yucca Mountain is undoubtedly the most thoroughly studied piece of real estate on the planet. Scientists, poking around the mountain since 1978, have spent billions of dollars just to make sure it is a geologically appropriate site. All the studies appear to support the suitability of the site.

Nevertheless, in a classic case of the NIMBY mindset, the Yucca Mountain project has been met with fierce opposition from many Nevadans. Some argue that it is unfair for nuclear waste to be stored in a state where there are no nuclear power plants, an objection that conveniently ignores the fact that Nevadans get about 15 percent of their electricity from a nuclear power plant in Arizona. Other Nevadans harbor a deep-seated resentment toward the federal government's ownership of 87 percent of Nevada's land and vent their frustration at the Yucca Mountain project. It is worth noting that Yucca Mountain lies within the Nevada Test Site where hundreds of nuclear weapons have been tested; therefore, it is an unlikely location for future hotels or casinos. Because Nevada has become an up-for-grabs "battleground state," the controversy has spilled over into presidential politics in the last few elections. Whether the Yucca Mountain Repository will ever actually open is still much in doubt.

## Visiting Information

Yucca Mountain lies in Nye County in southern Nevada, about one hundred miles northwest of Las Vegas. The Yucca Mountain Information Center is located at 2341 Postal Drive in Pahrump, Nevada, about a seventy-minute drive from Las Vegas. The Information Center has displays that address the questions of why Yucca Mountain was chosen, how the radioactive waste will be transported, and how different countries handle the waste problem. The Information Center is open Monday through Thursday from 9:00 A.M. until 4:00 P.M. As of this writing, the public tour program at Yucca Mountain has been suspended, owing to severe budget cuts. (The Information Center is also in danger

| Websites: | Information Center: www.ymp.gov/contact/ymsc.shtml Tours: www.ymp.gov/contact/tours.shtml |
| Telephone: | Information Center: 775–751–7480 Tours: 800–225–6972 |

of being closed.) The tours may be reinstated in the future if the budget allows. Check the website for an update. Past public bus tours included a stop at the top of Yucca Mountain for a talk on the geology and hydrology of the site and a visit to the south portal to see the tunnel boring machine. The tours last all day and include a lunch break. Advance registration is required.

# 8
# Chemistry in Industry

*. . . one or two atoms can convert a fuel to a poison,*
*change a color, render an inedible substance edible,*
*or replace a pungent odor with a fragrant one.*
*That changing a single atom can have such consequences*
*is the wonder of the chemical world.*

P. W. Atkins

Improvements in our health and standard of living are tightly coupled to progress in basic science. Nowhere is this relationship more evident than in chemistry. During the last two centuries, advances in our understanding of chemistry have given rise to the chemical industry and other related industries that turn raw materials into useful products. Applied chemistry has created medicines to improve our health, pesticides and fertilizers that increase the food supply, synthetic dies and fibers for the clothes we wear, synthetic sweeteners and flavors to make our food taste good, artificial rubber for our cars, celluloid for the movies, plastics for packaging, and soaps and cosmetics to keep us clean and good-looking.

The production of chemicals can be traced all the way back to 7,000 B.C., when Middle Eastern artisans refined alkali and limestone for glassmaking. During the sixth century B.C., the Phoenicians produced soap, and by the tenth century A.D., the Chinese were producing black powder for explosives. Large-scale chemical industries were born during the Industrial

Revolution. Among the first was British entrepreneur James Muspratt's mass production of soda ash (used for soap and glass) in 1823. Today, the chemical industry (defined as those companies that produce chemicals for various industries) is a $2-trillion global enterprise led by giant multinational corporations such as BASF, Dow, Shell, Bayer, INEOS, ExxonMobil, and DuPont. Nearly 80 percent of the industry's revenues are generated by plastics and polymers including polyethylene, polystyrene, and polyvinyl chloride (PVC). In the United States, the chemical industry is comprised of 170 major companies, generates $400 billion a year, and employs about a million people. The greatest concentration of the chemical industry's basic manufacturing plants is located along the coast of Louisiana and Texas, partly due to the area's proximity to key raw materials like petroleum. According to the American Chemistry Council, the top ten basic chemicals produced in the United States in the year 2000 (by weight, not revenue) were sulfuric acid, nitrogen, ethylene, oxygen, lime, ammonia, propylene, polyethylene, chlorine, and phosphoric acid.

Unfortunately for the scientific traveler, few major companies related to the chemical industry offer tours of their facilities. Reasons for this reluctance range from safety and security to budgetary constraints and proprietary concerns. The sites described below represent just a few industries where the science of chemistry plays a starring role. For more places to visit, check out the website at www.factorytoursusa.com.

## Borax Mine, Boron, California

Boron, the fifth element on the periodic table, does not exist on its own in nature. Instead, boron combines with oxygen and other elements to form boric acid or inorganic salts known as borates. The term "borax" refers to several closely related chemical compounds and minerals, all of which contain boron but vary in water content. Borax usually appears as a white powder of soft, colorless crystals that quickly dissolve in water, and borax is widely used in laundry detergents and other cleaning products because of its bleaching effects and stain-removing ability. In fact, stroll down the laundry detergent aisle at your local grocery store, and you will find a brand named "Borax." Other major commercial uses include fiberglass and heat-resistant glass, ceramic tiles and glazes, fertilizers, and insecticides. The most exotic use of boron is in nuclear reactors, where it is used to make the neutron-absorbing control rods.

Although trace amounts of borates exist in rocks, soil, and water, boron-containing ores are among the rarest minerals on earth. Concentrations of these ores, however, can be formed by repeated evaporation of seasonal lakes. The deposit found in the Mojave Desert was formed in this way between 12 and 18 million years ago. Discovered in 1925, it is one of the two major borate deposits in the world (the other is in Turkey). Owned by Rio Tinto, a leading mining company based in London, the Borax Mine supplies nearly half the world's demand for industrial borates.

## Visiting Information

To learn about the mine, make your way to the Borax Visitor Center near the appropriately named town of Boron, California, in the Mojave Desert north of Los Angeles. Begin your visit with a seventeen-minute video that relates the history of the mine and provides an introduction to the science of borates. After the film, take some time to wonder through the center's six exhibit areas. One display traces the geological history of the deposit and describes the mining operation, while another details how the ore is refined and distributed. The mine's safety and environmental programs are highlighted, and a display features a wide range of commercial products that use borates. A final exhibit tells the story of Borax in Death Valley during the late nineteenth century when teams of mules were used to haul away the ore. (For more about the history of the region's borate mining, visit the Twenty Mule Team Museum in Boron.)

Now go outside to the viewing area and cast your gaze on the big hole in the ground. A mile wide, a mile-and-a-half long, and 700 feet deep, this is the largest open-pit mine in California. For an even better view, climb the stairs to the roof of the visitor center. On display outside of the visitor center is one of the humongous 240-ton mining trucks that carry the ore out of the mine; for comparison, the car you drove to visit the mine probably weighs about two tons. Nearby is an area where visitors can collect their own mineral specimens. Be sure to claim your free sample of a mineral called ulexite. Better known as "TV Rock," its crystals act like tiny optical fibers that transfer the image of an object to the opposite surface. This property can be observed by placing the rock on a printed page and noticing that the print appears on the top surface of the rock.

Website: www.borax.com
Telephone: 760–762–7588

The visitor center is open daily, except major holidays, from 9:00 A.M. to 5:00 P.M. Admission is $2 per car. The scientific traveler may want to combine a visit here with a trip to NASA's Dryden Flight Research Center at Edwards Air Force Base. Also nearby in and around Barstow are some of the largest solar power installations in the world.

# Breweries

According to Benjamin Franklin, "Beer is proof that God loves us and wants us to be happy." Well, there must be a lot of happy people around because after water and tea, beer is the most consumed drink in the world. About 35 billion gallons of beer are sold worldwide every year generating annual global revenues of approximately $300 billion. The largest brewing company in the world is Anheuser-Busch InBev followed by SABMiller. Beer is also one of the world's oldest beverages, dating back to as early as 6,000 B.C. References to beer can be found in the written histories of ancient Egypt and Mesopotamia. The earliest direct chemical evidence of beer dates to between 3,500 and 3,100 B.C. at a site in western Iran. During the Middle Ages in Europe, poor sanitation made water unsafe to drink, and it was replaced by beer. Why include an entry on breweries in a chapter on chemicals in industry? Because, as we shall see, there's a lot of chemistry involved in brewing beer.

Beer is made from four simple ingredients: barley, a type of grain similar to wheat; hops, the cone-shaped flower of the hop vine, which is a member of the hemp family; yeast, a single-celled microorganism; and water. (Although barley is the preferred grain for making beer, any kind of grain can be used, including wheat, rice, oats, rye, and corn.) But brewing beer is not a simple matter of dumping the ingredients into a kettle and stirring; rather, brewing involves a complex series of biochemical reactions that take place under carefully controlled conditions of time, temperature, and other variables.

The first step is to coax the husk of the barley open so that the seed just begins to sprout or germinate. To accomplish this, the barley is soaked in hot water for several days. The water is drained off, and the damp barley is allowed to sit for several more days at about 60°F. In this process, called malting, the grain releases enzymes that can convert the starches contained within the seed into sugars that can sustain the plant until it begins photosynthesis. The trick is to stop the germination process at the point after the enzymes have been released but before they have had a chance to convert much of the starch into sugar. At this point, the malted barley is dried. Most commercial breweries buy barley that has already been malted according to their exact specifications.

Next comes a process called "mashing," which begins by crushing the malted barley to break up the kernels. The crushed grains are then sprayed with heated water and allowed to sit for an hour or two at a temperature of around 150°F. Now the enzymes go to work on the starches, which are long chains of glucose molecules. The enzymes break the long chains apart into short chains of only two or three glucose molecules. These short chains are fermentable sugars. The resulting liquid is drained from the bottom of the container and recirculated to the top where it filters through the husks of the grain—a process called "lautering." The grains are then washed with water to remove as much of the fermentable liquid as possible—a process called "sparging."

The resulting sweet and sticky liquid, called wort (pronounced wert), is poured into a huge brew kettle and brought to a vigorous rolling boil for about ninety minutes. Hops are added at various times during the boil to give the beer its bitterness, flavor, and aroma. The bitterness comes from acids, and the flavor and aroma comes from oils, both of which are released from the hops during the boiling. The boiling also gets rid of any enzymes left over from the mashing process. The "hopped wort" is then cooled to prepare it for the yeast.

To begin the fermentation process, the cooled wort is placed into a vessel containing the yeast. When the glucose molecules enter the yeast, they are broken down in a ten-step chemical process called glycolysis. The end products of glycolysis are two sugar molecules called pyruvates and some adenosine triphosphate (ATP), which gives the yeast the energy it needs to multiply. In the final step, the yeast converts the pyruvates into ethyl alcohol (ethanol) and carbon dioxide. The fermentation process requires from a week to a month, or more, depending on the type of yeast and the strength of the beer being brewed. For most beers, the alcohol concentration is between 4 and 6 percent by volume.

## Visiting Information

Numerous breweries across the country welcome visitors. Of those, I briefly describe three of the largest. All the tours conclude with free samples of beer (assuming, of course, you are age twenty-one or older).

The MillerCoors brewery in Golden, Colorado, is the largest single-site brewery in the world. Founded by Adolph Coors in 1873, this facility can brew up to 22 million barrels of beer each year. It claims to be the only major brewery in the United States that does most of its malting on-site. The free self-guided audio tours begin at the corner of Thirteenth and Ford streets and last thirty-five minutes. Tours are offered from 10:00 A.M. to 4:00 P.M. Thursday through Monday, except on Sunday when the tours begin at noon. The brewery is closed on Tuesdays and Wednesdays.

The MillerCoors Milwaukee brewery is another of the world's largest with a production capacity of ten million barrels of beer per year. Visitors can enjoy free one-hour guided walking tours that take you through the steps of the brewing process. The tours start at the Visitor Center at 4251 West State St. and end at the Bavarian-style Miller Inn. The tour includes a stop at the historic Caves, a restored section of the original brewery where beer was stored in the days before modern refrigeration. Tours are usually available from 10:30 A.M. to 3:30 P.M. from Monday through Saturday, but it is suggested that visitors call ahead to confirm tour times.

The Anheuser-Busch brewery in St. Louis offers tours that stop at the Budweiser Clydesdale paddock and stable, the lager cellar, the Brew House, and the Bevo packaging plant. Anheuser-Busch is the only major brewer that continues to use beechwood aging to age and naturally carbonate its beers. This is done by lining the fermentation tanks with a layer of beech wood chips. This provides a larger surface area for the yeast to settle

| Websites: | www.millercoors.com |
| | www.budweisertours.com |
| Telephone: | Golden, Colorado brewery: |
| | 866–812–2337 |
| | Milwaukee, Wisconsin brewery: |
| | 800–944-LITE |
| | St. Louis, Missouri brewery: |
| | 314–577–2626 |

on. The free seventy-five-minute tours begin at the Tour Center at Twelfth and Lynch streets. Tours are generally offered from 9:00 A.M. to 4:00 P.M. Monday through Saturday and from 11:30 A.M. to 4:00 P.M. on Sunday. The hours vary slightly through the year so check the website for exact times. Anheuser-Busch also offers tours of its breweries in Merrimack, New Hampshire; Jacksonville, Florida; Fort Collins, Colorado; and Fairfield, California.

## Distilleries

Beer is limited to a maximum alcohol content of about 15 percent by volume because most yeasts cannot reproduce in an environment where the alcohol concentration is any higher than that. At that point, fermentation stops, and no more alcohol is produced. But suppose you craved a beverage with a little more of a kick to it—something with a higher alcoholic content? To create these desirable drinks, fermented liquids like beer or wine are subjected to a process known as distillation. Distillation separates the chemical components of a liquid by heating it to the point of vaporization, cooling the vapor so that it condenses, and then collecting the condensate. This results in a purified or concentrated form of the original liquid. Distillation is a physical separation process; no chemical reactions are involved.

The apparatus used to perform distillation is called a still, of which there are two main types: a pot still and a continuous still. A pot still, the type associated with moonshiners of the prohibition era, has the form most people would recognize as a still. It consists of a copper or copper-lined pot with a spherical bottom and a long tapering neck. The neck is connected by a copper pipe to a condenser in the form of a cooled spiral tube. When a liquid mixture is heated in the pot to its boiling point, the composition of the vapor is different from that of the original liquid. When the vapor is cooled and condensed, the resulting liquid has a higher concentration of the component with the lower boiling point. A fermented liquid is mainly a mixture of water, which boils at 212°F, and alcohol, which boils at 173.3°F. Thus, when the liquid is boiled in a still, the liquid that condenses will have a higher concentration of alcohol than the original liquid. If an even higher concentration of alcohol is desired, the condensed liquid can be placed into another pot still to repeat the process. Successive distillations yield higher concentrations of alcohol. For example, when made in a pot still, cognac and Scotch whiskey are distilled twice, and Irish whiskey is distilled three times.

A continuous still consists of a tall vertical cylindrical column filled with a series of horizontal perforated plates placed at various heights within the column. Steam enters the still from the bottom while fermented liquid enters from the top. As the fermented liquid trickles down over the hot plates, it vaporizes and rises along with the steam. The rising steam and vapor interact with the falling liquid. This causes more of the liquid alcohol to vaporize and some of the steam to condense. The vapor is drawn into vents along the side of the column that direct it to a condenser and a collector. In this way, a continuous still works like the redistilling process with multiple pot stills. The advantage of a continuous still is that it allows an uninterrupted flow of incoming material and outgoing product, whereas the pot still works only in relatively small batches. Modern distilleries use continuous distillation because large quantities of alcoholic beverages can be produced. (Continuous distillation is also widely used throughout the chemical industry in areas such as oil refineries, natural gas processing, hydrocarbon solvents, and petrochemical production.)

Any alcoholic beverage in which the alcohol content has been increased by distillation is called a "spirit." Spirits include whiskey (distilled from fermented grain and aged in wood), brandy (distilled from wine), rum (distilled from sugarcane juice or molasses), vodka (distilled from grain, but not aged), and gin (distilled from grain, unaged, but flavored with juniper berries and other ingredients). Chemically speaking, spirits are complex liquids containing anywhere from 200 to 300 flavoring compounds, including carbonyl compounds, carboxylic acids and their esters, tannins, terpenes, pyridines, and, of course, alcohols.

The first evidence for distillation dates to around 2,000 B.C. in ancient Babylonia. Here, archaeologists have identified unusually shaped clay pots believed to have been used to extract small amounts of distilled alcohol for use in perfumes. Alcohol itself was distilled for the first time by the Persians sometime between 700 and 800 A.D. Distilled alcoholic beverages debuted in Europe during the 1100s in the form of medical elixirs concocted by alchemists. Today, the U.S. distilled spirits industry generates roughly $100 billion in economic activity and employs more than 600,000 people. Approximately 440 million gallons of distilled spirits were consumed in the United States in 2008.

### Visiting Information

Visitors are welcomed at numerous large and small distilleries across the United States. Of those, I offer brief descriptions of three that are particularly friendly to visitors.

The Jack Daniel's Distillery in Lynchburg, Tennessee, the oldest registered distillery in the country, has been designated as a National Historic Landmark. Jasper Newton (Jack) Daniel chose this spot because of the pure, iron-free, limestone water produced by Cave Spring. The tours include a stop at one of the aging houses where the whiskey sits for four years in barrels made of white oak. In the rick-yard, sugar-maple wood is burned to make charcoal. The whiskey is slowly filtered through ten feet of the charcoal to give it a more mellow taste. Other stops include the mash room, the copper stills, the Cave Spring, and Jack Daniel's original office. The distillery is open daily from 9:00 A.M. to 4:30 P.M. The free tours last about seventy-five minutes, and a new tour begins every fifteen minutes. Ironically, the distillery is located in a dry county; although it's evidently O.K. to make whiskey here, it's not O.K. to drink it.

The Maker's Mark Bourbon Distillery in Loretto, Kentucky, a National Historic Landmark, has been around since the mid-1800s. The tours take you through all the steps of the bourbon-making process. To be legally called "bourbon," the whiskey must be aged in barrels for two years. The free tours are offered every hour on the half-hour from 10:30 A.M. to 3:30 P.M. Monday through Saturday. From March through December, Sunday tours are offered at 1:30, 2:30, and 3:30 P.M. The distillery is closed on major holidays. Maker's Mark , located directly south of Louisville, is one of six bourbon distilleries on Kentucky's "Bourbon Trail." It turns out that 90 percent of all bourbon is made in Kentucky because the water is perfectly suited for it.

The Casa Bacardi Visitor Center is located on the grounds of Puerto Rico's Bacardi Rum Distillery, the largest rum distillery in the world. The rum is still produced with a secret strain of yeast that the company's founder, Don Facundo Bacardi, isolated in 1862. The free self-guided audio tour traces the history of the distillery and the steps involved in making the rum from sugarcane molasses. The Visitor Center is open Monday through Saturday from 8:30 A.M. to 5:30 P.M. (the last admission is at 4:15 P.M.) and on Sunday from 10:00 A.M. to 5:00 P.M. Casa Bacardi is located in Cataño, just outside San Juan.

> Websites: www.jackdaniels.com
> www.makersmark.com
> www.casabacardi.org
> Telephone: Jack Daniel's: 931–759–6180
> Maker's Mark: 270–865–2099
> Casa Bacardi: 787–788–1500

# Crane Museum of Papermaking, Dalton, Massachusetts

Paper is one of the most common and useful products in modern society. In fact, the average American consumes about seven hundred pounds of paper products every year. But just how, exactly, is paper made? The raw material for most papermaking is wood from trees. It takes about 6 percent of an average-size tree to make a ream (500 sheets) of paper. Wood is composed of cellulose fibers held together by lignin along with sugars and other organic compounds. The fibers are the key ingredient used for papermaking so the first task is to separate the cellulose from the lignin and other chemicals. This pulping process results in a material called, appropriately enough, wood pulp. Two main methods exist for making this pulp: mechanical and chemical.

The interior of the Crane Museum of Papermaking.

In mechanical pulping, the separation of the cellulose is accomplished simply by finely grinding or chopping the wood. In this way, 90 percent or more of the wood can be turned into pulp—a very efficient process. The disadvantages are twofold: the pulp still contains the lignin that can turn the paper yellow or brown when exposed to sunlight, and the short, stiff fibers produce paper that isn't very strong. Thus, mechanically produced pulps are typically used for newspaper, toilet paper, and other low-strength applications.

In chemical pulping, heat, pressure, and chemical reactions are used to dissolve the lignin away from the cellulose. The most common method of doing this is called the "kraft" process, where the wood and chemicals are cooked in a "digester" that removes 90 to 95 percent of the lignin. (This cooking of the wood produces much of the characteristic stench associated with paper mills.) Kraft mills account for about 75 percent of all pulp made in the United States. This method preserves the length of the fibers, which results in stronger paper, and the digester's waste can generate much of the heat and electricity needed to run the process. Only 40 to 50 percent of the original wood, however, ends up as paper.

Next, the pulp is highly diluted with water. The watery mixture is sprayed onto a mesh screen, which allows the water to escape but traps the fibers on the screen. The spraying is repeated over and over again to form mats that are fed between rollers to compress and dry the material. The mats, sometimes as wide as ten yards, are laid out in continuous rolls that can be as long as thirty miles. Finally, the mats are cut into paper of the desired size. The most commonly used size for office paper, 8.5 by 11 inches, originates from the fact that the screens that held the pulp in the early days of papermaking were framed in standard-sized boxes that measured 44 inches long and 17 inches wide; cutting twice along the length gave 11 inches and a single cut along the width gave 8.5 inches.

The earliest forms of paper were made from the papyrus plant in ancient Egypt in about 3500 B.C. The origins of the modern papermaking process can be traced back to A.D. 105 in China, where a court official named Ts'ai Lun, produced paper from a slurry of mulberry fibers. Paper remained expensive until the advent of steam-powered papermaking machines in the 1800s. Mass-produced paper made books and newspapers cheaper and affordable. Pencils and the invention of a practical fountain pen meant that people now had the tools they needed to write.

## Visiting Information

Perhaps the best place to learn about papermaking in the United States is the Crane Museum of Papermaking in western Massachusetts. The Crane family has been in the papermaking business since 1770, when Stephen Crane took over the first paper mill in Massachusetts, proudly proclaiming his revolutionary spirit by calling it The Liberty Paper Mill. In 1801, Stephen's youngest son, Zenas, found a perfect spot for a new mill on the banks of the Housatonic River and founded the family company. The major difference between high-quality Crane paper and regular paper is that Crane paper is made from cotton rather than wood. The cotton is in the form of clippings from garment factories. Water from the river was used to wash the cotton rags and, in the early days, to power the mill's machinery.

Since 1879, Crane & Co. has supplied the U.S. Treasury with paper for use in the nation's currency. Paper money in the United States is the most durable in the world: a $1 bill has a circulation life span of eighteen months (higher denominations last longer). Crane & Co. is an industry leader in deterring counterfeiting through the sophisticated use of security threads and fibers, watermarks, special additives, and fluorescent and phosphorescent elements. The company continually seeks out new technology in its never-ending fight to stay a step ahead of the counterfeiters.

The Crane museum is housed in Crane's Old Stone Mill, a building dating from 1844 complete with rough-hewn oak beams, colonial chandeliers, and a ceiling resembling the inverted hull of a ship. Begin your visit by viewing the twenty-minute video that describes the company's history and shows what goes on behind the factory's walls. (There are no tours of the factory itself.) Exhibits trace the history of papermaking in America to the Revolutionary War. A scale model of part of the original mill shows how paper was made—one sheet at a time—back in the early 1800s. You can see some original hand molds that Zenas Crane used from 1801 to 1831. Be sure to admire the collection of distinctive Crane papers

> Website: www.crane.com
> Telephone: 413–684–6481

used for currency, stock and bond certificates, and the elegant stationery of movie stars and presidents. And don't miss the disposable paper shirt collars, fashionable during the post–Civil War era.

The museum is open, free of charge, from early June through mid-October. Hours are Monday through Friday from 1:00 P.M. until 5:00 P.M. Guided tours of the museum can be arranged by calling at least a week in advance.

# Glass-making

As I write, I am sitting snugly by my fireplace looking out through the window at my backyard. It is mid-December, and the wind chill temperature is –20°F. The glass provides me with an undistorted view of the outside world while shielding me from the harsh winter environment. What exactly is this magical material that lets in the light but keeps out the cold and blocks the wind?

The most common type of glass—the kind found in windows, bottles, and jars—is mostly silicon dioxide with a little soda and lime thrown in to the mix. Silicon dioxide, more commonly known as silica, makes up much of the Earth's crust and is found in quartz, quartz sand, and sandstone. Soda (sodium oxide) is added to lower the melting temperature while the lime (calcium oxide) makes the glass insoluble in water and much more durable. The mixture is heated in a large ceramic vessel to a temperature of approximately 3,000°F. The ingredients melt together forming a viscous fluid that flows like thick syrup. This molten glass is cooled to a working temperature of about 1600°F, and globs of it are injected into molds. Air is blown into the molds forcing the glass to the outer walls. The glass is slowly cooled until it solidifies into a bottle or a jar.

While making glass containers is fairly straightforward, producing a piece of glass with perfectly flat surfaces for use in windows is a surprisingly difficult task. (Surfaces that are not perfectly flat result in distortion when looking through the glass.) It is done by pouring the glass onto a pool of molten tin. Because the tin is much denser than the glass, the glass floats on top of the tin without mixing. The top surface of the liquid glass forms a perfectly flat surface naturally. The bottom surface of the glass conforms to the top surface of the tin, which is also perfectly flat. This "float" method of making windowpanes was developed in 1959 and revolutionized window-making. It can produce flat glass ranging in thickness from less than a millimeter to several centimeters.

Chemically, glass is classified as an amorphous solid, a material with no long-range order to the atomic structure. (This is in contrast to a crystal, where the atoms are arranged in a regular, repeating pattern.) If you were shrunk to subatomic size and placed on any atom in a glass, you could predict what kind of atoms your nearest neighbors are, but if you travel more than two or three atomic diameters from your starting point, you could not make such a prediction.

The use of glass goes back to our prehistoric ancestors who used a naturally occurring glass called obsidian, a shiny rock produced in volcanic eruptions. It can appear black, orange, gray, or green in color. Obsidian can be shaped to produce the sharp, durable edges needed for stone tools. The first glass was manufactured around 4,000 B.C. in Egypt and Mesopotamia. This early glass, simply heated crushed quartz, was used as a glaze for ceramic objects. The discovery of glass was probably accidental—perhaps someone placed some sand in an oven used for ceramics and noticed that a clear liquid was produced.

## Visiting Information

The best place to learn about the history, science, and technology of glass is the Corning Museum of Glass in Corning, New York. The museum is home to a world-class collection that exceeds 45,000 objects spanning 3,500 years of glass-making history. In the museum's Optics Gallery, you can see the 200-inch diameter glass disk that was the first attempt at making a mirror for the Hale telescope on Palomar Mountain. The Vessels Gallery tells the story of all types of glass containers from simple bottles and jars to complex television picture tubes and missile nose cones. Visit the Windows Gallery to learn more about how windows are made. In addition to exhibits, the museum offers thirty-five live shows and demonstrations every day. Among these are a flame-working demo where glass rods are transformed into fanciful figures, a glass-breaking demo where you can observe how different types of glass shatter, and a "Magic of Glass" show where you can enjoy demonstrations that illustrate the unique properties of glass. For an extra charge, you can even make your own glass souvenir. The museum's hours are 9:00 A.M. to 5:00 P.M. daily and it is open until 8:00 P.M. in the summer from Memorial Day through Labor Day. The museum is closed on major holidays. Admission is $12.50 for adults with discounts for seniors, students, and AAA members. The museum is free for ages 19 and under. Advance reservations are recommended for the glass souvenir making experience. See the website for details.

Another interesting place to witness the glass-making process is Kokomo Opalescent Glass in Kokomo, Indiana. Founded in 1888, the company claims to be the oldest maker of opalescent glass in the world. Opalescent glass is milky in appearance and can be made in a variety of colors. It is a decorative glass used for such items as stained-glass win-

*Courtesy of the Corning Museum of Glass*

A youngster blowing his own glass souvenir at the Corning Museum of Glass.

dows and Tiffany lampshades. The folks in Kokomo can manufacture more than 22,000 different color, density, and texture combinations. The one-hour tours lead you through the glass-making process, starting at a giant round furnace with a dozen openings into the glowing interior. Inside each opening sits a ceramic pot that can hold a thousand pounds of glass. The ingredients for a particular type of glass are carefully measured and mixed. The batch is then dumped into the pots where it melts at temperatures exceeding 2400°F for seventeen hours. Next, the molten glass is ladled out and deposited onto a mixing table where glass from the different batches is mixed together. Finally, the glass is pressed into a sheet before it is cut and made ready for shipping. Tours are offered on Wednesdays and Fridays beginning promptly at 10:00 A.M. The factory is closed on holidays and during the month of December. Admission is $1.

Websites: Corning Museum of Glass:
www.cmog-px.rtrk.com
Kokomo Opalescent Glass:
www.kog.com

Telephone: Corning Museum of Glass:
888–310–1249
Kokomo Opalescent Glass:
765–457–1829

# Hagley Museum and Library (DuPont Company), Wilmington, Delaware

DuPont is currently the world's second largest chemical company (behind BASF) with 60,000 employees in more than seventy countries. The company's wide-ranging businesses are organized around five areas: electronic and communications technologies, performance materials, coatings and color technologies, safety and protection, and agriculture and nutrition. DuPont built two of the very first research and development laboratories in the United States and today operates more than forty laboratories in this country and eleven more abroad. DuPont is a model of how investment in basic science can yield practical and profitable results. The company's logo includes the words: "The Miracles of Science." It is a fitting phrase, for few companies have performed more scientific miracles than DuPont.

The company was founded in 1802 by Eleuthère Irénée (E. I.) du Pont, the youngest son of a Paris watchmaker. At age fourteen, du Pont wrote a scientific paper on the manufacture of gunpowder and went on to study the chemistry of explosives alongside the great French chemist Antoine Lavoisier. The du Pont family's moderate political philosophy did not mesh well with the revolutionary spirit in late eighteenth-century France. In 1797, an angry mob looted du Pont's shop, and the family was thrown in jail for a short time but escaped the chaos by jumping aboard a ship and sailing to America, where they arrived in January 1800.

It didn't take long for du Pont to discover that the U.S. gunpowder industry lagged well behind that of Europe. Seizing the opportunity to build a business, he rushed back to France in 1801 to buy the latest powder-making equipment and raise capital. Upon returning to America, he chose a location on the banks of the Brandywine River near Wilmington, Delaware, to build his first powder mill. The site was convenient to all the resources he needed to run the mill: water to power the machinery, timber and willow trees to provide the high-quality charcoal needed for the powder, granite quarries for building materials for the mill, and proximity to the Delaware River for shipping. After he established the site, the DuPont company grew rapidly and eventually became the largest black powder manufacturer in the world. During the Civil War, DuPont made half of the gunpowder used by the Union Army.

In the early 1900s, the company shifted its focus from explosives to chemicals and materials. DuPont built a research laboratory called the

Machinery used to harness the power of water for the original powder mill.

"Experimental Station" across the river from the original powder mills and the breakthroughs started coming: Fabrikoid, an artificial leather; Duco, a quick-drying finish; and a moisture-proof cellophane wrap are but a few of the materials developed at the lab in the early days. In 1927, DuPont hired Dr. Wallace Carothers, a Harvard organic chemist, to do basic scientific research on polymers, very long molecules with a repeating chemical structure analogous to the links of a chain. Carothers proceeded to discover neoprene, the first synthetic rubber, and nylon, the first truly synthetic fiber. The company used nylon to make women's hosiery. The product went on sale to the general public in May 1940, and women across the country lined up at department stores to buy the stockings, which eventually became known simply as "nylons." Other synthetic materials developed by DuPont include Teflon for nonstick cookware, Mylar for film and audio and video tape, and Kevlar for bullet-proof vests. Freon, a chemical used in refrigeration and air conditioning, was invented by two General Motors scientists, but Freon was manufactured by DuPont. When researchers discovered that Freon causes ozone depletion, DuPont worked to create environmentally benign alternatives.

DuPont's chemical and material innovations have been recognized with four National Medals of Technology and a Nobel Prize. DuPont chemist Charles Pederson won the 1987 Nobel Prize in chemistry for his discovery of crown ethers. In recent years, the company has announced that it will begin trying to produce chemicals from living plants rather than processing them from petroleum. Also, the company has moved to incorporate the field of biology into its scientific research.

## Visiting Information

The rich history of DuPont is preserved at the Hagley Museum and Library, located on the site of the original gunpowder works. The museum features exhibits that trace the company's development from an explosives manufacturer to a modern research-based firm that has changed the way we live. Here, you can try on a space suit made from eleven layers of material including Nomex, Kevlar, Dacron, and nylon—all DuPont products. Explanations of the scientific concepts behind the products accompany many company items on display. NASCAR fans enjoy sitting in Jeff Gordan's #24 Chevrolet Impala. DuPont has been sponsoring Gordan since 1992 in an effort to keep DuPont brands and products in the public eye. A separate exhibit area explores how simple machines help make life easier.

The du Pont family home at the Hagley Museum.

Elsewhere on the grounds is the Powder Yard where you can see how water was used to run the machinery during the company's early days. Docents can demonstrate the workings of a powder tester, a steam engine, and a water turbine. At Workers Hill, you can experience what life was like for the powder mill workers and their families. The Brandywine Manufacturer's Sunday School, built in 1817, is where children came to learn to read and write before public education was established.

Finally, Eleutherian Mills is the name given to the du Pont family's first residence in America, built by E. I. du Pont in 1803. Five generations of du Ponts have lived in the home, furnished with family antiques and memorabilia. Next to the house is a restored French-style garden similar to the one planted by E. I. du Pont himself. Also near the house is the barn that holds a Conestoga wagon and other vehicles

> Website:  www.hagley.lib.de.us
> Telephone:  302–658–2400

along with a collection of agricultural equipment and weather vanes. On the barn's lower level you can find an assortment of antique cars from the early 1900s—a period when DuPont was closely associated with General Motors. The car collection includes a 1911 Detroit Electric car and a 1928 DuPont Motor Phantom. Another exhibit near the home is the company's first office. Built in 1837, it remained the center of operations for more than fifty years. The office is furnished with typewriters, ledgers, and a telegraph key from the period.

# Hillestad Pharmaceuticals, Woodruff, Wisconsin

Vitamins, complex organic molecules that, in very small amounts, serve a wide variety of functions in the human body, are essential for good health. Some assist in mediating the body's chemical reactions while others function as hormones or antioxidants. Vitamins are usually designated with a letter, such as vitamin C. For historical reasons, several vitamins were lumped together under the label of vitamin B. These are now distinguished from one another by the addition of numbers as in vitamin $B_1$ (thiamine) or vitamin $B_2$ (riboflavin). With the exception of vitamin D, which can be produced through the action of sunlight on the skin, all vitamins must be taken in with the food that we eat. The B vitamins along with vitamin C dissolve in water and hence are not retained by the body. Your body's supply of these vitamins must be replenished every day. In contrast, Vitamins A, D, E, and

K, known as fat-soluble vitamins, can be stored in the body. Absorbing too much or too little of these vitamins can result in health problems. A deficiency in vitamin C causes a degeneration of tissues resulting in bleeding from the gums, poor wound healing, and eventual death. This disease, called scurvy, was common among sailors on long voyages due to a lack of fresh fruits and vegetables. As a remedy, the British Royal Navy supplied their sailors with lemons and limes, good sources of vitamin C; this earned the British sailors the nickname "limeys." Similarly, the bone disease rickets is due to a vitamin D deficiency. These diseases are rare in the developed world, given its adequate food supply and the addition of vitamins and minerals to foods, a process called "fortifying." In addition to ingesting vitamins through the food we eat, inexpensive pills containing recommended doses of vitamins are taken every day by millions of Americans. Scientific evidence supports the benefits of dietary supplements for some health conditions but not for many others. In the United States, the advertising for dietary supplements must include a disclaimer that the product is not intended to treat, diagnose, mitigate, prevent, or cure disease and that any health claims made by the manufacturer have not been evaluated by the Food and Drug Administration. Dietary supplements are considered to be foods, not drugs. Because of this, vitamin manufacturing is not regulated to the same high standards as medical pharmaceuticals.

## Visiting Information

One company that makes these so-called "dietary supplements" is Hillestad Pharmaceuticals. Their products are made from natural ingredients, and the company keeps more than 5,000 different raw materials in its inventory ready for use. Hillestad is not a mainstream pharmaceutical company in the mold of Eli Lilly or Pfizer. Unlike the major pharmaceutical companies, Hillestad welcomes visitors to its factory where you can see how they make millions of tablets every day.

Tours begin with the first step in tablet making: weighing the ingredients and granulating them in a blender. Water is added to the dry ingredients allowing them to be blended together. A low temperature dryer dries out the material to form a powder. Next, the powder is compressed into tablet form by punches that can exert up to 10,000 pounds of pressure per square inch. The tablets are then coated to make them more durable, eliminate odors or aftertaste, and to make them smoother and therefore easier to swallow. Coating is accomplished in two different ways. One method uses a row of spinning stainless steel coating pans that can each hold up to 60,000 tablets. As the pans spin, a coating is added, and the tumbling tablets coat each other. The other method involves a high-speed coating machine that can coat a quarter of a million tablets in an hour. The coated tablets are now ready for packaging. Machines count out the tablets and dump them into waiting bottles. In a final step, the bottles are labeled, coded, and sealed.

Of course, to do you any good, a tablet's chemicals must be absorbed by your body. For that to happen, the tablet must break up after being ingested. The last stop on the tour is a sort of artificial stomach where the time it takes a tablet to break up (a property called "disintegration time") is measured. This is done using a pan containing the acids found in a human stomach. A tablet is moved through the acids to simulate the process of digestion. Batches of tablets that don't disintegrate in the proper time are rejected.

Hillestad Pharmaceuticals is open to visitors Monday through Friday from 9:00 A.M. to 3:00 P.M. Tours last

> Website:  www.hillestadlabs.com
> Telephone:  800–535–7742

about half an hour, require no reservations, and are free. A neighboring outlet store sells the company's many products. In addition to vitamins and minerals, the company sells protein pills, skin care products, and natural remedies.

## Mary Kay Cosmetics, Dallas, Texas

Women have been applying cosmetics to enhance their appearance and aroma for thousands of years. The first solid archaeological evidence of the use of cosmetics dates from around 4,000 B.C. in Egypt. The ancient Egyptians were a very spiritual people and equated their level of spirituality to their appearance: better looking implied a higher spirituality. One item in the Egyptian cosmetics case was a powder called mesdemet made from a mix of copper and lead ore. A typical application might place green mesdemet on the lower eyelids while black or dark grey was brushed on the upper eyelid and eyelashes. In addition to improving one's appearance, mesdemet was an excellent disinfectant and insect repellent. Another Egyptian cosmetic concoction was something called kohl, a dark powder made of a mixture of burnt almonds, oxidized copper, lead, ash, ocher, and different colors of copper ore. Kohl was applied with a small stick to the eye area forming an almond shape. In addition, the Egyptians used red clay mixed with water to color their lips and cheeks, and they extracted a dye from the leaves of the henna plant to color their nails orange or yellow.

The Greeks adopted many Egyptian cosmetic practices and products, but they weren't interested in looking good for the gods, but rather for each other. Later, the Romans used fat from sheep mixed with blood for nail polish and took baths in mud seasoned with crocodile feces. The Roman playwright Plautus once said that "a woman without paint is like food without salt."

In Europe, a pale face became a fashion statement that endured for centuries. In the sixth century, women achieved the pale look by taking the extraordinary step of bleeding themselves. In the Middle Ages, paleness was a sign of wealth because the rich did not have to toil away in the fields under a blazing sun that resulted in tanned and rugged skin. Less extreme measures were taken during the Renaissance, when women used a whitening agent composed of carbonate, hydroxide, and lead oxide to give their faces a porcelainlike appearance. Unfortunately, repeated applications of this noxious mixture resulted in an array of serious health problems, including muscle paralysis and even death. Not until the nineteenth century was poisonous lead oxide replaced with harmless zinc oxide, a chemical that continues to be used in facial powder today. The lead-related deaths were unintentional, but in one of the most notorious episodes in cosmetic history, Italy's Signora Toffana purposefully used arsenic in a face powder marketed toward wealthy women. When their husbands died from ingesting the poisonous powder, the women inherited the riches. Toffana was executed after her powder killed 600 unsuspecting husbands.

In Elizabethan England, women applied egg whites to their faces to create a glazed look. They also bleached their hair with lye which eventually caused their hair to fall out so wigs were introduced. But the heavy, elaborate wigs had to be held in place with animal lard, which attracted lice and other nasty little pests. Oh, the things people endure just to look pretty! Cosmetics went out of fashion for a while in the 1800s during the reign of England's Queen Victoria, who advocated a strict commitment to morals and modesty among women. Victoria declared that makeup was improper and vulgar, relegating it to use only by actors and prostitutes. Nevertheless, some Victorians continued to use cosmetics, a few of which contained poisonous ingredients. For example, lead and antimony sulfide was used for eye shadow, mercuric sulfide was used for lip reddener, and belladonna (deadly nightshade) was used for sparkling powder. Makeup made a comeback in the late 1800s. The pale look became passé, and a more natural look was in vogue.

By the middle of the twentieth century, cosmetics were widely used throughout the developed world. To avoid the sometimes poisonous mixtures of the past, most countries thoroughly test and highly regulate the ingredients in modern cosmetics. Testing cosmetic products on animals is a controversy swirling around the industry. Animal testing has been banned across the European Union. Today, cosmetics generate $230 billion in global sales; L'Oréal is the largest cosmetics company in the world followed most notably by Revlon, Estée Lauder, Proctor and Gamble, and Tokyo-based Shiseido. But the only company to offer public tours is Mary Kay Cosmetics.

After being passed over for a promotion in favor of a man whom she had trained, Mary Kay Ash decided to start her own company. So, in 1963, Ash, along with her son Richard Rogers and $5,000 in savings, started a company called "Beauty by Mary Kay." Operating out of a storefront in Dallas, the company grew rapidly, especially after Ash was interviewed on *60 Minutes* in 1979. Mary Kay products are not sold in stores; they are available directly from "Independent Beauty Consultants." Top sales people are famously awarded pink Cadillacs. Today, the company makes more than two hundred beauty products, has a sales force of 1.8 million consultants, and generates about $2 billion in sales.

## Visiting Information

Mary Kay Cosmetics offers tours of both its manufacturing plant and its corporate headquarters building, but the manufacturing plant tour is of more interest to the scientific traveler. On this tour, you see several of the forty or so production lines where a combination of machine and manpower prepare the items for shipping. On the fragrance line, watch as bottles are filled by machine with perfume, sprayers are inserted by hand, lids are machine tightened, and finally hand-packed in pretty boxes. This probably ranks as the best smelling assembly line in the world! On the eye shadow line, plastic gray bottoms with shallow wells are filled with a rainbow of colors, then coupled with clear tops. Approximately 38,000 kits are produced here every ten-hour shift. Elsewhere, a variety of creams and foundations are mixed in giant vats. As empty bottles parade by on conveyor belts, metal nozzles dance up and down like carousel horses, squirting their creamy contents into the containers.

The highlight of the tour is the lipstick line, where liquid lipstick is pumped into cylindrical metal molds. The molds are then placed on an ice-cold cooling table to allow the liquid to solidify. Each "bullet" of lipstick is manually inserted into its tube. The conveyor belt moves at an impressive speed, and yet the workers somehow manage to keep up. (The scene is reminiscent of an episode of *I Love Lucy* where Lucy and Ethyl fight a losing battle trying to wrap pieces of candy along a rapidly moving conveyor belt.) In a final step, the tubes are fed into a glass-enclosed oven, where flames lick the outer layers, giving the lipstick its sheen.

Guided tours of the factory are free, but require reservations at least three days in advance. The one-hour tours are scheduled for 2:00 P.M. on Mondays, 10:30 A.M. on Fridays, and both 10:30 A.M. and 2:00 P.M. on Tuesdays through Thursdays. The Mary Kay Museum, located on the lobby floor of the corporate headquarters, can be visited without reservations anytime Monday through Friday from 9:00 A.M. to 4:00 P.M. Here, you can watch a video that tells the Mary Kay story, see artifacts from the company's history, and admire evening gowns worn by Mary Kay.

> Website: www.marykay.com
> Telephone: 972–687–5720

Guided tours of the headquarters are also available and include a visit to Mary Kay's personal office along with the museum. The factory is located at 1330 Regal Row in Dallas and the corporate headquarters is at 16251 Dallas Parkway in Addison, just north of Dallas.

## 3M Museum, Two Harbors, Minnesota

Scotch Tape, Post-it Notes, and sandpaper are but a few of myriad products made by 3M, originally known as the Minnesota Mining and Manufacturing Company. In fact, the company's products are so commonplace that 3M claims that a quarter of the world's population uses at least one 3M product every day. The company was founded in 1902 by a group of five investors who believed a local deposit of the mineral corundum had been discovered. Corundum was used as an abrasive for grinding wheels and sandpaper, and the investors' idea was to mine and sell the substance to manufacturers on the east coast. But there was a big problem with the plan: the rocks they thought were valuable corundum turned out to be the worthless mineral anorthosite. The company teetered on the precipice of failure until one founder convinced Lucius P. Ordway, a St. Paul businessman, to invest in the enterprise. With the influx of capital, the company converted an abandoned flour mill into a sandpaper factory. After years of struggling to produce good quality sandpaper, the company paid its first dividend of 6 cents a share in 1916. That same year, the company's general manager bought a laboratory for $500 and encouraged the employees to experiment; 3M became a science-based company.

The first major breakthrough came in the early 1920s when 3M developed the world's first waterproof sandpaper. The product greatly improved the quality of automobile finishing and reduced the airborne lead-based paint dust that threatened the health of workers. The company's close ties with the auto industry prompted work on pressure-sensitive tape, and in 1925 a young lab assistant named Richard G. Drew invented masking tape. A few years later, cellophane tape was developed. The "Scotch" brand name is derived from a complaint made by a dissatisfied customer who told a 3M salesman to "Take this tape back to your stingy Scotch bosses and tell them to put more adhesive on it." The name "Scotch" stuck (no pun intended). The decade of the 1940s saw the development of Scotchlite Reflective Sheeting for highway signs, magnetic recording tape, and offset printing plates; with the 1950s came Thermo-Fax copying, Scotchgard Fabric Protector, and videotape; and the 1960s ushered in dry-silver microfilm and overhead projectors.

The advent of one of the company's simplest and most ubiquitous products, Post-it notes, can be traced back to 1968, when 3M scientist Spencer Silver created a "low-tack," reusable pressure sensitive adhesive. But not until 1974 did Art Fry, a colleague of Silver, come up with a use for the new adhesive. Fry, a member of his church choir, was annoyed that the slips of paper he used to mark his hymnal kept falling out. As his mind wandered during a sermon, it occurred to Fry that the adhesive might be used to anchor his bookmarks. The product was launched in 1977, but failed, evidently because consumers were unfamiliar with how the sticky pieces of paper could be used. The company tried again a year later, but this time it gave free samples to the residents of Boise, Idaho, so they could try the product. The product took hold, and by 1981 Post-it notes were being sold across North America and Europe.

Today, 3M tinkers around with various technologies including adhesives, abrasives, light management, microreplication, nonwoven materials, nanotechnology, and surface modification. The company has a presence in more that sixty countries and employs more than 75,000 people worldwide, 7,000 of whom work in research and development. Recent products include optical films for LCD televisions, Scotch Transparent Duct Tape, and Post-it Super Sticky notes with improved adherence to vertical surfaces.

## Visiting Information

This green two-story building, where the articles of incorporation for 3M were signed, served as the company's original office and headquarters. In 1991, the Lake County Historical Society bought the building and, with generous financial help from 3M, restored, and renovated, and turned it into a museum. Exhibits include a recreation of the original office, displays containing photographs and original artifacts and documents that tell the company's history, a lab area highlighting 3M's dedication to research, and a hands-on area to help visitors understand the underlying technology. A wall of holograms holds 3D images of 3M products. Timelines trace the development of Scotch Tape and Post-it notes. Nearby, a giant roll of sandpaper along with a sandpaper mosaic pays tribute to the company's first product.

The 3M Museum is located at 203 Waterfront Drive in Two Harbors, Minnesota, about twenty-five miles northeast of Duluth. The museum is open from May through October. Hours are Monday through Friday from 12:30 P.M. to 5:00 P.M., Saturday from 9:00 A.M. to 5:00 P.M., and Sunday from 10:00 A.M. to 4:00 P.M. Admission is $2.50 for adults and $1.00 for youth ages 9 to 17.

> **Website:** www.lakecountyhistoricalsociety.org
> **Telephone:** 218–834–4898

## Tom's of Maine, Sanford, Maine

If you prefer products made from all-natural ingredients based on plants and herbs, Tom's of Maine may have just what you're looking for. Tom's uses hops to control odor in its deodorant, pure flavor oils in its toothpaste, and a mixture of Vitamin C and rosemary as a preservative in its soap. No artificial colors, flavors, sweeteners, preservatives, or animal-based ingredients are used in the company's line of products. It all started in 1968, when Tom and Kate Chappell decided to leave the hustle and bustle of urban life in Philadelphia and "move back to the land" in rural Kennebunk, Maine. Unable to find natural personal care items, they took matters into their own hands and made their own. The ingredients for Tom's first batch of homemade toothpaste were mixed in a Kitchen Aid blender.

In 1970, financed by a $5000 loan from a friend, the entrepreneurial couple went into business. The company's first product was the country's first nonphosphate liquid laundry detergent sold in containers labeled with a postage-paid mailer so that customers could return them to be refilled and reused. Today, Tom's of Maine employs roughly 200 workers making ninety

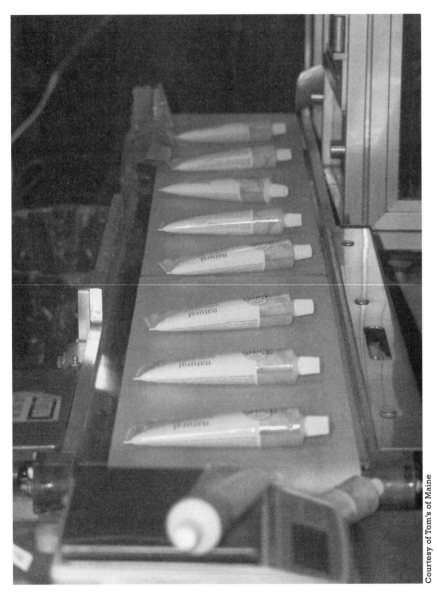

Toothpaste coming off the production line at Tom's of Maine.

Courtesy of Tom's of Maine

personal care products ranging from toothpaste and deodorant to soap, shampoo, and shaving cream. The company's natural fluoride toothpaste is the only natural toothpaste to win the American Dental Association's Seal of Acceptance. In 2006, Tom's of Maine became an independently run division of the Colgate-Palmolive Company.

Tom's of Maine prides itself on being environmentally friendly. To this end, the company uses energy-efficient fluorescent lighting in its facilities, recycles everything from cardboard to shrink wrap, and buys energy credits from a wind farm in Nebraska. The company even gives employees $4,000 if they purchase a new hybrid car. The company is also "animal-friendly" by avoiding the use of ingredients derived from animals and by pioneering innovative alternatives to animal testing.

## Visiting Information

Tours of Tom's of Maine begin at Kate's Herb Garden where you can inspect some of the natural ingredients used in the company's products. Next, you'll have to don a lovely hair net before going inside the manufacturing area to see how toothpaste is made. The toothpaste is mixed in giant, 3,000 pound vats, pumped into recyclable aluminum tubes, and packaged using biodegradable shrink wrap. The machinery produces toothpaste at the impressive rate of eighty tubes per minute. Workers on the packaging line switch tasks every hour to combat the boredom that comes with a repetitive job.

The Tom's of Maine factory is located in Sanford, Maine. The forty-five-minute tours, offered only during the summer from mid-June through Labor Day, are free, but reservations must be made by calling the number to the right. Guests must wear closed-toed shoes. The tour is not recommended

> Website: www.tomsofmaine.com
> Telephone: 800–775–2388

for children younger than five, and the facility is not handicap accessible. A short video of the tour is available on the website.

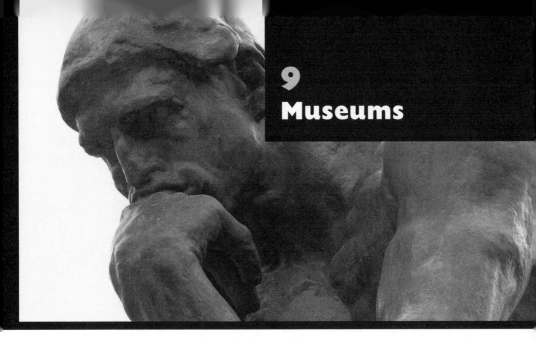

*No one ever flunked a museum.*　　　　　　　　　　　　Frank Oppenheimer

The word "museum" comes from Greek words meaning "a place for the Muses," the nine Greek goddesses who presided over the arts, literature, and the sciences. Science museums are places that can excite the senses and engage the intellect. Relax . . . there are no tests or quizzes to take. After a visit to science museum, you may notice things in the world around you that you haven't noticed before. You may even be inspired to read a book or take a class. Since the 1960s, the number of science museums and centers worldwide has exploded. Approximately 350 science museums in the United States alone generate a combined attendance numbering in the tens of millions.

The origins of today's science museums can be traced to the Renaissance, when wealthy aristocrats collected rare or unusual objects to show to guests and when universities began maintaining specimen collections to aid in the training of students. The first true science museum, the Museo de Ciencias Naturales (Museum of Natural Science) in Madrid, opened in 1752. During the Industrial Revolution, national exhibitions were held to celebrate advances in science and technology. These temporary exhibitions sometimes evolved into permanent museums. London's Science Museum, for example, was a by-product of the Great Exhibition in the Crystal Palace in 1851. In America, local or regional natural history societies began collections that morphed into museums. Perhaps the oldest science museum in the United States was the New England Museum of Natural History, established in 1864. It is now known as the Boston Museum of Science.

Until a few decades ago, the most commonly heard phrase in a science museum was: "Do not touch!" Museums were filled with static objects to look at and admire from afar. Barriers of glass and rope separated the visitor from the exhibits. This began to change in the early twentieth century when Munich's Deutches Museum pioneered the idea of interactive exhibits that invited you to push a button, pull a handle, or turn a crank. This radical notion made its way to the United States when Julius Rosenwald, chairman of Sears, Roebuck, visited the Deutches Museum and decided to build a similar museum in Chicago. As a result of Rosenwald's encounter, the Museum of Science and Industry opened its doors in 1933.

The concept of an interactive science museum reached a new level when the San Francisco Exploratorium opened in 1969. The Exploratorium, the brainchild of physicist Frank Oppenheimer (the brother of J. Robert Oppenheimer) who created exhibits that demonstrated phenomena in new and interesting ways, soon became the model for other science museums and freely shared its innovative exhibit designs with other museums.

In the early 1970s, giant IMAX (an acronym for Image Maximum) theaters featuring an image several stories high, made their debut. In a variation of IMAX, called OMNIMAX or IMAX Dome, the viewer is surrounded by the image, giving the viewer the sensation of actually being part of the scene. The first OMNIMAX Theater opened as part of the Reuben H. Fleet Science Center in San Diego in 1973. Today's so-called "science centers" usually include an Exploratorium-style science museum, an IMAX theater, and a planetarium.

In this chapter, I have included four major science museums that feature a strong physics/chemistry component. I have also included three smaller, lesser known museums that specialize in physics or chemistry. Happy musings!

## Exploratorium, San Francisco, California

The Exploratorium, advertised as a museum of science, art, and human perception, was once identified by *Scientific American* magazine as the best science museum in the world. I, for one, would not argue with that assessment. Its unusual and truly innovative exhibits have been copied by museums

The Palace of Fine Arts in San Francisco, location of the Exploratorium.

throughout the world, a practice that the Exploratorium encourages by publishing detailed exhibit design plans. As a professor at the University of Colorado, Frank Oppenheimer became convinced that students learn best through active experimentation. He threw out the standard physics course and replaced it with a "library of experiments" where students, guided by their own interests, could learn at their own pace. His success with this approach, coupled with visits to European museums, persuaded Oppenheimer that America needed a new type of science museum—one that would bridge the gulf between the complexity of science and technology and the daily life and experience of the average person.

After a long search, Oppenheimer finally found a home for his museum at the Palace of Fine Arts in San Francisco, a cavernous building with three acres of open space that had been built for the Pacific International Exposition of 1915. The museum opened its doors in 1969 with only a few dozen mostly borrowed exhibits. Oppenheimer immediately established a machine shop equipped for carpentry, electronics, and welding, where new exhibits could be designed, built, and repaired. Located near the entrance, the shop is enclosed by glass so that visitors can see the work going on inside. As a way of field testing an exhibit, prototype displays are placed just outside the shop for people to play with.

Oppenheimer died in 1985, but his spirit lives on through hundreds of exhibits spanning a wide range of topics, including biology, light and color, vision, music, language, hearing, patterns, memory, and electricity and magnetism. Many exhibits were created by visual and performing artists working in conjunction with scientists and educators. The exhibits are really activities that provide instructions and ask you to notice something. An explanation of the observed phenomena follows. Thus, the Exploratorium exhibits force you to do and think rather than passively watch.

In keeping with the theme of human perception, one of the oldest and most popular exhibits is the "Tactile Dome." Here, enveloped in total darkness, you are forced to creep and crawl your way through an anthill-like maze of various materials and shapes. According to the original press release, "The idea is to make people aware of what a complex, sensitive, and underused sense touch is, and to train them to use the astonishing range of its perceptions, which include detection of pressure, pain, temperature and kinesthesia, as well as internal body and muscle awareness." This unforgettable experience requires an advance reservation and a separate admission fee.

Within walking distance of the Exploratorium is the "Wave Organ," a wave-activated acoustic sculpture situated on a jetty that forms the small boat harbor in the Marina District. This installation, sponsored by the Exploratorium, consists of twenty-five organ pipes made from PVC pipe and concrete located at different elevations to account for high and low tides. Sounds are produced by waves hitting the ends of the pipe and the resulting motion of water in and out of the pipes.

## Visiting Information

The Exploratorium is located in San Francisco's Marina District at 3601 Lyon Street. To get here via public transportation, take bus #28, 29, 30, or 43 or the Golden Gate Transit bus. The museum is open Tuesday through Sunday from 10:00 A.M. until 5:00 P.M. It is closed on Mondays. Admission is $14 for adults, $11 for students, and $9 for children ages 4

Website: www.exploratorium.edu
Telephone: 415–563–7337

through 12. The Tactile Dome is $17 with general admission included. The Exploratorium is included on the CityPass, which gives you discounted admission to five San Francisco attractions plus a seven-day bus pass. Free parking is available.

# Franklin Institute, Philadelphia, Pennsylvania

Founded in 1824 as the first professional organization of mechanical engineers and draftsmen in the United States, the Franklin Institute is now composed of three centers: The Center for Innovation in Science Learning (CISL), The Benjamin Franklin Center, and The Science Center. The CISL provides services to school teachers and administrators such as programs that encourage the use of technology in the classroom and on-line lessons and activities. The Benjamin Franklin Center runs The Franklin Awards Program, one of the world's oldest and most prestigious science and technology awards program, recognizing outstanding achievement in chemistry, computer and cognitive science, earth science, engineering, life science, and physics. Many Franklin Medal winners have gone on to win the Nobel Prize. The Franklin Center also operates the Benjamin Franklin National Memorial, one of the few national memorials located outside of Washington, D.C. The public face of the Institute is the Science Center which includes the museum, the Fels Planetarium (the second oldest planetarium in the United States), and the Tuttleman IMAX Theater. The institute has seen its share of history through the years. Here in 1893 Nikola Tesla demonstrated the principle of the wireless telegraph, and in 1934 Philo T. Farnsworth gave the first public demonstration of an all-electronic television system. On a less auspicious note, on March 31, 1940, a Franklin Institute employee issued a press release stating that institute astronomers had confirmed that the world would come to an end the next day at exactly 3:00 P.M. After the resulting panic had subsided, the institute's misguided April Fooler was dismissed.

The museum's Franklin Gallery is the place to see some of the institute's most prized possessions related to Benjamin Franklin. A portion of Ben's original lightning rod system is on display, along with reproductions of a pair of bifocals and a Franklin stove. You can test your skill at using a "long-reach arm" used by Franklin to grab books from the top shelf of his library. Have you ever rubbed a wet finger around the edge of a wine glass making it "sing"? This is the idea behind Franklin's glass armonica, a musical instrument consisting of rotating glass bowls of various sizes. The instrument is played by pressing a wet finger onto the glass.

The Franklin Gallery leads into the "Electricity" area filled with all kinds of electrical gadgets and gizmos accompanied by demonstrations that illustrate basic electrical principles. Among the devices are a collection of early light bulbs, a primitive movie projector, and a scale model of Morse's original telegraph. A set of four unique and amusing automata teach safety lessons about electricity. In one, a cat knocks a radio into a bathtub occupied by his unfortunate master. Nearby is one of the museum's oldest displays: a diorama of Franklin's kite experiment.

More of the museum's early exhibits can be seen in the adjoining "Wonderland of Science" area, which highlights the role the institute played in major scientific breakthroughs. (The phrase "wonderland of science" was used to describe the museum shortly after its opening.) Don't miss the displays on early television and motion picture technology including an odd-looking mechanical (as opposed to electronic) television receiver. The first photograph of lightning can be seen nearby along with a poster advertising the world's first electrical exposition, organized by the Institute in 1887. No trip to the Franklin Institute would be complete without visiting the Giant Heart in the neighboring exhibit hall. A museum icon since it first appeared in 1954, this two-story heart is the right size for a 220-foot-tall person. A walk through this model traces the path of blood through the heart.

Tucked away in a corner of the second floor is the "Franklin Air Show," an exhibit dedicated to the science and technology of flight. Many interactive stations demonstrate the basic physical principles of flying such as air pressure, lift, drag, and angle of attack. Don't be shy about donning the foam wings on your arms, standing in front of a giant fan, and feeling the lifting force as you angle your wings into the wind. Sprinkled among the demonstrations are some truly amazing relics from the Wright Brothers. The Franklin Institute was the first scientific organization to formally recognize the Wright Brothers as the first men to achieve powered flight. Orville Wright rewarded the Institute by donating a large collection of artifacts from his workshop. Among the Wright-related items are airfoils (including the one chosen for the 1903 flyer), automatic stabilizers, a balance to measure the drag force, and a concept drawing of the 1903 flyer. Look up to see the centerpiece of the exhibit, an actual 1911 Wright Model B Flyer, one of the first aircraft to be mass produced. The plane has been restored to the condition of its first flight including muslin covered wings and a working engine.

The newest continuing exhibit at the Franklin Institute, called "Amazing Machine," dissects machines and explains their basic mechanical components like pulleys, gears, cams, and linkages. The exhibit reveals the mysterious inner workings of everyday household machines, such as vacuum cleaners, power drills, and thermostats. You can even test your skill at operating a real backhoe or watch a giant can crusher flatten a garbage can.

A display on time-keeping has reproductions of delicate clock movements designed by Galileo and Robert Hooke along with a six-foot tall model of the Strasbourg cathedral clock. But the star of this exhibit, built in 1810 by the Swiss clockmaker Henri Maillardet, is an automaton that can actually draw four images and write three poems. In contrast to the electronic programming that controls modern robots, this automaton's smooth motions result from brass disks with hills and valleys along the edge. A unifying theme of the exhibit is that engineering is an art form, a theme manifested in three large kinetic art installations by sculptor and architect Ben Trautman.

On the third floor is "Sir Isaac's Loft," a room full of fun physics demonstrations. One of the more unusual items here is an artistic interpretation of a nuclear chain reaction. Small lights arranged in a rectangular array mimic the motion of atomic nuclei. When one light meets another light, one of the lights splits into two lights and those two lights meet other lights producing four lights and so on. Also of interest is a model of a perpetual motion machine designed by Charles Redhoeffer in 1812. The term "perpetual motion machine" generally refers to a device that supposedly produces more energy than it consumes, a blatant violation of the well-established law of conservation of energy, which states that energy can be neither created nor destroyed. Scientists and engineers sneer at such devices and yet attempts to build such machines persist to this day. This particular machine is one of the most famous and fascinating of these fakes. Redhoeffer exhibited this machine in Philadelphia and charged admission to see it work. An argument erupted regarding the authenticity of the machine, and large bets were placed on the outcome. Finally, Coleman Sellers, a young boy who happened to be a mechanical genius, inspected the machine and figured out how it really worked.

The Benjamin Franklin National Memorial, a twenty-foot-tall marble statue of Franklin seated in a chair, is located in the rotunda of the Franklin Institute. The statue is surrounded by a collection of Franklin's personal possessions including his composing table and several original publications. For more on Benjamin Franklin and related sites, see the entry in chapter 1.

### Visiting Information

The Franklin Institute is located at 222 North Twentieth Street (at the intersection of Twentieth Street and Benjamin Franklin Parkway) in Philadelphia. A parking garage is located behind the building at Twenty-first and Winter Street. Have your parking ticket validated for a reduced rate. The

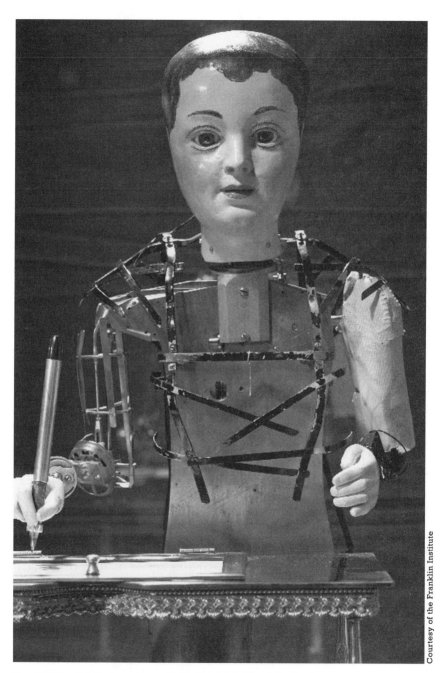

The Maillardet Automaton at Philadelphia's Franklin Institute.

base admission rate for adults is $14.25, which includes a planetarium show. There are reduced rates for seniors, students, military personnel, and children. Teachers with an ID get in free! The museum is open daily from 9:30 A.M. until 5:00 P.M.

> Website: www.fi.edu
> Telephone: 215–448–1200

## Museum of Science, Boston, Massachusetts

The Museum of Science traces its origins back to 1830, when six gentlemen with a shared interest in the natural world formed the Boston Society of Natural History. The society displayed its specimens, mostly stuffed animals, in a series of temporary facilities until 1864, when the men purchased a building in Boston's Back Bay and promptly dubbed it the New England Museum of Natural History. The museum remained there until after World War II, when a ninety-nine-year lease was signed with the city of Boston for a parcel of land spanning the Charles River Basin. The museum pays the state the exorbitant sum of $1 a year for use of the land, now known as Science Park. Opened in 1951, the new museum today holds more than five hundred exhibits along with a planetarium and a domed Omnimax theater and attracts over 1.6 million visitors annually.

The museum's defining attraction and a definite "must see before you die" for any serious scientific traveler is the Theater of Electricity, home to the world's largest air-insulated Van de Graaff generator. (The reader may remember the Van de Graaff generator from physics class—it's the machine used to make a student's hair stand on its end.) Whereas most Van de Graaff generators used for physics demonstrations stand about three feet tall, this behemoth stands two stories tall and can create about 2.5 million volts of electricity. The machine was built by Dr. Robert Van de Graaff himself, a professor at nearby MIT, who originally designed the generator for use as a particle accelerator. For this purpose, one sphere held a positive charge, the other, a negative charge. The spheres were connected with a tube containing a target. When the charge was large enough, a five-million-volt discharge occurred between the two spheres, hitting the target. Each dome held a laboratory with equipment that could be used to study what happened when the target was smashed. Some years later, the two spheres were joined to form one large terminal for high-energy X-ray experiments. The right column held the working parts of the machine, while the left column contained the equipment that created the X-rays. In the early 1950s, after the machine's experimental usefulness was exhausted, MIT donated the generator to the museum. Today, the left column is empty and functions only to support the sphere.

In the Theater of Electricity, the Van de Graaff is accompanied by two big Tesla coils measuring about eight feet tall and four feet wide across the top. Tesla coils are basically air-cored transformers, devices that can change or transform low voltages to high voltages or vice-versa. The museum's twin Tesla coils can jack the voltage up to 250,000 volts and pump it out into the air in the form of giant sparks of electricity.

The Van de Graaff generator and the Tesla coils play a starring role in the half-hour demonstration shows that take place at least twice daily in the Theater of Electricity. The show begins with a short introductory talk and a revving up of the Van de Graaff to throw a few sparks (just to warm up the audience a bit). A re-creation of the most famous electrical experiment ever performed—Benjamin Franklin's kite experiment—is next on the agenda. There is some question about whether Franklin actually performed this experiment, but he certainly did not do it the way it is often portrayed. (See the entry on Benjamin Franklin in chapter 1.) In this reenactment, without wind to fly the kite with a string, the kite hangs from a plastic pole. A string from the kite leads down into a jar where a key is attached. One or two inches below the key is a grounded metal ball. As the demonstration proceeds, you notice that the tail of the kite is drawn toward the dome of the Van de Graaff. This happens because the negative charge on the Van de Graaff dome rearranges the charges within the atoms of the paper tail, a process called polarization. Because opposite charges attract and like charges repel, the negatively charged electrons are pushed away from the dome leaving the positively charged nuclei just a tad closer to the dome. Because the electrical force varies inversely with the square of the distance between the charges (for example, if the distance between the charges is doubled, the force is one-quarter as much), the attractive force between the negatively charged dome and the positive charges in the paper tail is stronger than the repulsive force between the dome and the negative charges in the tail. Closeness wins, and the tail is attracted to the dome.

Another effect that may be visible with keen eyesight is a faint purple glow around the edges of the kite. This coronal glow is known as Saint Elmo's fire, named after the patron saint of sailors, who reported seeing the glow around the masts of ships as they sailed through a storm. The phenomenon happens because the electricity strips the electrons off (ionizes) the surrounding air molecules and creates a plasma. When the electrons recombine with the nuclei, the light emitted is visible as a glow.

A second demonstration illustrates why your car is a safe place to be during a lightning storm. It is not, as many believe, because the car's rubber tires insulate you from the ground. As is pointed out during the show, air is also an insulator; if lightning can travel through two or three miles of air, then it can certainly travel through a couple of inches of rubber with no trouble. In fact, it would take a piece of rubber a mile thick to protect you from lightning. To demonstrate the correct reason, the brave lecturer steps into a metal cage and is lifted toward the dome of the Van de Graaff. When a spark hits the cage, the person inside is safe and uninjured, even when he or she touches the inside of the cage. How is this possible? A common, but incorrect, explanation is that because your car (or in this case, the cage), is made of metal, an electrical conductor, the charges can move around as they please. Because like charges repel, they get as far away from each other as possible (that is, on the outside of the cage). This is called a "Faraday Cage." But this explanation holds true only for static electricity, and bolts of lightning are not static. The correct explanation involves something called the "skin effect." Lightning is extremely high frequency alternating current. For complicated reasons involving what are called eddy currents, the current in alternating current tends to flow on the outside or "skin" of the conductor. If the demonstrator were to touch the outside of the conductor, as does accidentally happen on occasion, he or she would receive a painful, though not fatal, shock. In the show's grand finale, all the machines are charged up to create an indoor lightning storm with sparks flying in every direction. Dramatic music adds to the charged atmosphere. It is a sight the scientific traveler will not soon forget!

The best exhibition at the Museum of Science is "Mathematica: A World of Numbers and Beyond" created by the famous husband-and-wife design team of Charles and Ray Eames. This survey of mathematics' greatest hits explains the most compelling ideas from algebra, geometry, calculus, probability, topology, Boolean algebra, and logic. Here, you can witness the spontaneous generation of a bell curve as thousands of balls fall through a grid of hundreds of pegs, landing in slots to form the familiar pattern. The "Image Wall" reveals the mathematical patterns hidden in the natural world: sunflower seeds follow the Fibonacci series of numbers, and the shell of a chambered nautilus forms a Golden Spiral. The "History Wall" traces the development of mathematical ideas from 1100 A.D. through to 1950. Watch the closed loops of wire of various shapes dip into soapy water, pulling out to show the minimal surface for that shape. Elsewhere, sand pours through a swinging bucket to form a Lissajous figure. Math fanatics and phobics alike will enjoy this classic exhibit.

There is, of course, much more to see in the Museum of Science. Physics-related highlights include the fascinating optical illusions in the "Seeing is Deceiving" area and an extensive display of machine parts inviting you to press a button and watch them work. The "Light House" has optics displays, and the "Take a Closer Look" area has a cloud chamber and several other interesting demonstrations.

### Visiting Information

The Museum of Science is located in Science Park in downtown Boston. To get there, take the Green Line subway to Science Park. The museum is open daily from 9:00 A.M. until 5:00 P.M. On Fridays, the museum stays open until 9:00 P.M. and hours are extended during most of the summer. Admission to the exhibit halls is

> Website: www.mos.org
> Telephone: 617–723–2500

$19.00 for adults, $17.00 for seniors age 60 and over, and $16.00 for children ages 3 through 11. There is an extra charge for other attractions.

## Museum of Science and Industry, Chicago, Illinois

In 1911, Julius Rosenwald, then president of Sears, Roebuck and Company, vacationed in Germany, visited the Deutches Museum in Munich, and returned to Chicago, where he embarked on a mission to establish an American center for "industrial enlightenment" and science education. Rosenwald began by pledging $3 million to convert the Palace of Fine Arts, the only surviving building from the 1893 World's Fair, into an interactive science museum. Following Rosenwald's lead, other business leaders opened their wallets and contributed to the cause. The Museum of Science and Industry opened its doors to the public in 1933, and today it attracts approximately two million visitors annually, making it one of the top seven most visited museums in the United States. It is the largest science museum housed in a single building in the Western Hemisphere with exhibits spanning three levels: the balcony, the main floor, and the ground floor. We'll

Chicago's cavernous Museum of Science and Industry.

start at the balcony, where most of the basic physics and chemistry exhibits are located, and work our way down.

The Regenstein Hall of Chemistry contains exhibits on biochemistry, food, polymers, acids and bases, and elements, compounds, and molecules. Look for the red circles marked "Touch Me." Your touch initiates chemical reactions such as mixing chemicals to produce light or electricity as in a battery. The best exhibit in this hall is a giant Periodic Table of the Elements along the wall. Samples of most of the elements are displayed along with a list of their uses. With a push of a button, a recording of the Tom Lehrer song "The Elements" will set the mood for your chemical contemplations.

In the Grainger Hall of Basic Science, the nature of science is illuminated by panels on the controversy over creationism vs. evolution and the contrast between astrology and astronomy. Other panels discuss the use of the word "theory" in science, compare science to philosophy, religion, and the arts, and point out that scientific research is a group effort. The development of ideas about the nature of light is used as a case study of how science progresses. The "History Wall" traces important events in the fields of physics, chemistry, and biology from the year 1400. Be on the look-out for the "Finding Out" panel, where you see a rather unremarkable looking black pot. This is one of the hidden treasures of the museum: Robert Millikan's oil drop apparatus that he used to measure the electrical charge on the electron (see the entry on Robert Millikan in chapter 1). Another unusual object is a 100,000 watt incandescent lamp built in 1930. This lamp, equivalent to nearly 1700 60 Watt bulbs, could light up a room with the intensity of the noonday Sun in June. Nearby, a six-foot-tall Tesla coil built in 1935 can be seen along with a model of an early steam engine invented by Thomas Newcomen in England. Newcomen's design was later improved upon by James Watt. Machines like this one introduced the Industrial Revolution. Hands-on exhibits demonstrate the magnetic field around a wire and the generation of electricity by moving a magnet in and out of a coil of wire. A large cylindrical vat contains ping-pong-ball-sized spheres that can be sent whizzing around with a blast of air, mimicking the motion of molecules in a gas. The physics exhibits are interrupted by a fascinating display tracking the development of the human fetus.

Continue your walk and find the display addressing the question "How did life begin?" Here, you can examine the apparatus from the lab of Cyril Ponnamperuma, one of the world's leading researchers in this field. In the glass vat, the primordial ingredients of the Earth's primordial atmosphere are mixed and subjected to an electrical spark. After allowing the experiment to run for several hours, a thick brown liquid coats the inside of the glass. The liquid contains complex organic molecules, including amino acids. The first experiment of this type was done by Stanley Miller and Nobel laureate Harold Urey at the University of Chicago. The results of this type of experiment prove that the basic building blocks of life are easy to produce.

Further down the hall an exhibit on microscopes includes a replica of an early and surprisingly small microscope constructed by the Dutch tradesman Antonie van Leeuwenhoek. Leeuwenhoek's "microscopes" were really just powerful magnifying glasses consisting of a single lens made from small glass spheres. Although they could provide a magnification of up to 275x, Leeuwenhoek's microscopes were nothing like compound microscopes that use two lenses.

It's impossible to miss the gigantic ten-foot-tall Wimshurst machine, a mechanical device used to create high voltages. The bulk of the machine consists of two large vertically mounted discs that can be made to rotate in opposite directions using a hand crank. The machine on display here, built by Wimshurst himself in 1884, is the largest one he ever made. A smaller version that you can try out is mounted on the wall. Another impressive piece of apparatus is the tank from a cyclotron (a type of particle accelerator—see the introduction to chapter 5) that was used by physicists at the University of Michigan. Nearby, a television screen shows a simulation of the movement of an electron around a nucleus and how this leads to the idea of an electron "cloud." Finally, don't miss the classic science film "Powers of 10" by Charles and Ray Eames. This nine-minute film starts one meter away from a picnicker in a Chicago park and takes you on a journey first into outer space to the edge of the observable universe and then into inner space down to the protons and neutrons in an atomic nucleus. Another section of the balcony focuses on flight and houses a 2003 reproduction of the original Wright Flyer.

Moving down to the Main Floor, you may want to stop by the Whispering Gallery where one can whisper at the focus of one reflector and have a friend hear you at the focus of a second reflector at the other end of the room. In the next room an infrared camera detects the heat radiation emanating from your warm body. Try placing your hand on your clothing for a few seconds and then removing it. You'll see a handprint showing where the heat from your hand has warmed your clothing. Along the same hallway a new exhibit called "Earth Revealed" is definitely worth a look. Suspended from the ceiling of a circular theater is a six-foot diameter sphere upon which images of the Earth and other worlds can be projected. This realistic, three-dimensional view is nothing short of spectacular. Before leaving the Main Floor, stand in the rotunda and read the inscription: "Science Discerns the Laws of Nature. Industry Applies Them to the Needs of Man."

Continue down to the ground floor for a visit to the Energy Lab. The world's first nuclear reactor was built about a mile from the museum (see the Nuclear Sculpture entry in chapter 6) so it is not surprising to find an emphasis here on nuclear energy and radiation. Full-size replicas of a spent fuel container and a fuel assembly are on display along with a model of the first nuclear reactor assembly. A variety of panels and interactive exhibits explain how electricity is generated. Of historical interest is an actual DC (Direct Current) Dynamo built by Thomas Edison's electric company in 1886 along with a model of Edison's Pearl Street Generating Station in Manhattan. Edison's DC system lost out to the AC (Alternating Current) system devised by Nikola Tesla because AC could deliver electrical power over much longer distances.

The Henry Crown Space Center is where you'll find the single most precious artifact in the entire museum: the actual flown *Apollo 8* Command Module, the first manned spacecraft to fly beyond Earth orbit and circle the Moon. As the *Apollo 8* astronauts—James Lovell, Frank Borman, and William

Anders—orbited the Moon on Christmas Eve 1968, one of the most tumul-tuous years in American history since the Civil War, the crew read from the book of Genesis as television viewers watched the Earth rise above the barren lunar horizon. The images put our Earthly troubles into perspective. Another flown spacecraft is the *Mercury Aurora 7* capsule, the ship that carried Scott Carpenter into orbit. An actual Apollo Lunar Module Trainer is on exhibit surrounded by a collection of supplies and equipment and a Moon rock returned by the *Apollo 17* mission. Models of the next generation of NASA spacecraft, the Ares rockets, are on display near the entrance to the center.

Two famous exhibits in the museum are the Coal Mine, where you descend 50 feet underground to explore the inner workings of a mine, and the *U-505*, a German submarine that was captured in 1944. In 2005, the *U-505* was moved into an impressive new subterranean exhibition hall replete with newsreel footage and displays that firmly place the sub into its histor-ical context. A fifteen-minute tour of the sub's interior is well worth the extra $5 charge.

## Visiting Information

The Museum of Science and Industry is located at Fifty-seventh Street and Lake Shore Drive, about five miles from Millennium Park and downtown Chicago. To get here from downtown using public transportation, take the Metra Electric train from Millennium Station on Randolph Street to the 55–56–57th Street stop, but be sure to check the train schedule beforehand so you won't have a long wait. Another option is the #10 Museum of Sci-ence and Industry bus, which runs every day during the summer and on weekends the rest of the year. A taxi ride will cost you about $17 from Mil-lennium Park. The museum website lists more options. Parking at the museum costs $14.

The museum is open every day except Christmas. Regular museum hours are 9:30 A.M. until 4:00 P.M. Monday through Friday and 11:00 A.M. until 4:00 P.M. on Sunday. The museum is open until 5:30 P.M. during the summer from late May through June, July, and August. Admission is $13 for adults, $9 for children age 3 through 11, and $12 for seniors age 65 and older.

Website:  www.msichicago.org
Telephone:  1–800–468–6674

There is an extra charge for the OMNIMAX Theater. A number of free days are offered throughout the year. The Coal Mine and the U-505 exhibits attract long lines so head for these attractions early.

## Chemical Heritage Foundation Museum, Philadelphia, Pennsylvania

The Chemical Heritage Foundation, an organization dedicated to preserving and explaining the history of chemistry, has amassed one of the world's largest collections of artifacts related to the chemical and molecular sciences and industries. The collection includes works of art, photographs, documents, laboratory apparatus, and scientific instruments. Hundreds of these items are on display in the "Making Modernity" exhibit in foundation's new museum. This permanent exhibit tells the story of how chemistry has shaped the modern world, a saga that begins with glassmaking in the Roman Empire and ends with the silicon computer chips of today. Exhibit topics range from alchemy, electrochemistry, and synthetic materials to chemical education and chemistry sets. The instruments on display are accompanied by explanations of their significance. For example, an oxygen meter designed by Linus Pauling for use in World War II submarines was later adapted to measure oxygen levels in neonatal incubators (it turns out that some premature babies were being blinded by retinal damage caused by overexposure to oxygen). The museum also features temporary exhibits that change several times each year. The exhibits in the Fisher Gallery on the fourth floor can be viewed by appointment only.

### Visiting Information

The museum is located in the heart of Philadelphia's Independence National Historic Park at 315 Chestnut Street. Museum hours are Monday through Friday from 10:00 A.M. until 4:00

Website: www.chemheritage.org
Telephone: 215–925–2222

P.M. The museum is closed on most public holidays. Admission is free. Audio tours of the "Making Modernity" exhibit are available by cell phone. Ask for details when you arrive.

## Collection of Historical Scientific Instruments, Harvard University, Cambridge, Massachusetts

This collection of beautiful old scientific instruments is an absolute "must see" for any scientific traveler interested in the history of science and technology. Many instruments are nothing less than works of art in brass and wood. Formally begun in 1948, the Harvard collection now exceeds 20,000 objects dating from about 1400 A.D. to the present. A small fraction of the collection is on display here with instruments spanning a range of scientific disciplines from astronomy to zoology. Benjamin Franklin himself obtained some instruments on display. In 1764, a fire swept through Harvard Hall incinerating much of the university's scientific apparatus. Harvard deputized Franklin and assigned him the task of procuring state-of-the-art equipment to replace the instruments lost in the fire. Franklin visited his favorite London instrument makers and hand-picked items to be shipped back to Harvard. The items Franklin purchased on his European spending spree are marked with a bust of old Ben.

The Putnam Gallery on the first floor of Harvard's Science Center holds the permanent exhibits arranged in rough chronological order from sundials to a cyclotron, while the Special Exhibitions Gallery directly upstairs contains temporary displays developed by students, faculty, and invited scholars. A booklet is available upon entering the Putnam Gallery that describes, in general terms, the various areas of the collection. Reading the booklet as you move through the exhibits enhances your visit. Each instrument is identified with a number that corresponds to a key visible in the display case.

The first object you see as you enter the gallery is a spectacular Grand Orrery, a mechanical model of the solar system dating from 1788. The celestial dome is supported by bronze figures of Benjamin Franklin, Isaac Newton, and James Bowdoin, the Governor of Massachusetts. The figures were cast by a Boston silversmith, whose name you may recognize—Paul Revere. The case next to the Orrery contains all manner of instruments used for navigating, surveying, calculating, and time-keeping. Be sure to locate the geometrical and military compass made to Galileo's specifications by a Paduan artisan in 1603. The compass was used to aim cannon and to calculate the amount of powder needed.

Next is an exhibit of seventeenth- and eighteenth-century instruments such as microscopes, air pumps, barometers, and thermometers. Check out the "thunder house," a device for dramatically demonstrating the advantages of lightning rods. The house contains a "powder bomb" enclosing a small amount of gunpowder. When an electric spark is introduced to the house without a lightning rod, the spark is directed to the gunpowder causing a small explosion. Also interesting is the portable electric machine used to entertain people in taverns and coffeehouses.

The displays continue with eighteenth- and nineteenth-century astronomical instruments including telescopes (the earliest surviving example of a Gregorian telescope is on display) and equipment used for spectral analysis. Keep an eye out for Annie Jump Cannon's research notebook, which she used to develop a new system for classifying stars. Further down the gallery is a fascinating group of devices used by psychologists to study how we perceive and experience sound and light. Devices used to measure reaction time, an important part of early experimental psychological research, are also on display.

The gallery ends with a group of instruments used for twentieth-century physics research. On the wall is the cupola from a shed used by Harvard physicists to observe cosmic rays. A cloud chamber and counters accompany the cupola. Display cases tell the story of Harvard's wartime research into acoustics in an attempt to control sound on the battlefield. Loud noise during combat had a negative effect on the performance of fighter pilots and tank crews. Finally, relics from the Harvard Cyclotron Laboratory are shown including the cyclotron control console, preserved exactly as it was on the final day of operation, right down to the take-out menu from a Chinese restaurant.

## Visiting Information

The Harvard Collection of Historical Scientific Instruments is located in Room 136 (Putnam Gallery) and Room 251 (Special Exhibitions Gallery) of the Science Center on the Harvard University Campus in Cambridge, Massachusetts. Take the Red Line subway to the Harvard. The Putnam Gallery is open weekdays from 11:00 A.M. until 4:00 P.M. The Special Ex-

Website:  www.fas.harvard.edu/~hsdept/chsi.html
Telephone:  617–495–2779

hibitions Gallery is open weekdays from 9:00 P.M. until 5:00 P.M. Summer hours may vary, and the collection is closed on university holidays. There is no admission charge.

# The MIT Museum, Cambridge, Massachusetts

Founded in 1971, the mission of the MIT Museum is "to engage the wider community with MIT's science, technology, and other areas of scholarship in ways that will best serve the nation and the world in the 21st century." The museum accomplishes this mission in sensational fashion with exhibits that unite art and science, capturing the essence of MIT as a place where new ideas come alive. Exhibit topics include robotics and artificial intelligence, holography, stroboscopic photography, and the institute's history. The first floor of the museum is occupied by the "Innovation Gallery," where temporary exhibits showcase some of the latest technology from MIT's engineers. On my visit, the "City Car," a new concept in urban public transportation was on display. The tiny two-person cars would be made available to city-dwellers in stacks. When your trip is complete, just park it for the next customer to use. Nearby, a series of aquariums is home to mutated zebrafish used to study the differences between diseased and normal cell tissues. In the back of the gallery, several Remotely Operated Vehicles (ROVs) used to explore the deep ocean environment can be inspected. Suspended from the ceiling is a blue "Jason Junior," the vehicle that was used to discover the Titanic. The ROVs must be built to withstand the immense pressures in the depths of the ocean. These pressures are nicely demonstrated by a set of Styrofoam cups that have been submitted to various high pressures. The Innovation Gallery is interesting, but the museum's real treasures await you upstairs.

The first exhibit you encounter, "Robots and Beyond: Artificial Intelligence at MIT," explores the area of engineering the public most closely associates with the institute. Here, you will see the Minsky Arm, a robotic arm developed by robotic pioneers Marvin Minsky and Seymour Pappert, the Black Falcon, an arm that extends a surgeon's hands and fingers, and UNIROO, a one-legged robot that can hop up and down on one leg while keeping its balance. Don't miss the little antlike robots that are programmed to form social groups and interact cooperatively. The most famous robots you encounter here are COG, an anthropomorphic robot that possesses tactile and visual sensors that allow it to interact with its environment, and KISMET, a robot with a humanoid head complete with a mouth, ears, and large eyes complemented by animated eyebrows. KISMET was built to communicate with humans by voice and facial expressions.

In the "Holography: The Light Fantastic" exhibit, you can enjoy twenty-three historic holograms chosen from MIT's assortment of 1,800 pieces, the world's largest holography collection. One hologram invites you to look at the moon through a telescope. My favorite shows an attractive woman blowing a kiss as you scan across the image. Holography has practical applications. The "Lindow Man" hologram shows the remains of a man who got stuck in an English bog two thousand years ago, demonstrating

Justin Knight, Courtesy of the MIT Museum, Cambridge, Mass.

The "Robots and Beyond" exhibit at the MIT Museum.

holography's use in anthropology. Another hologram shows how architects can use the method to visualize a structure in three dimensions.

The mesmerizing kinetic sculptures of Arthur Ganson, the creator of the popular construction toy "Toobers and Zots," are next. These wondrously imaginative contraptions are made from simple materials ranging from metal wires to chicken bones and are driven by basic electric motors or hand cranks. In one gizmo, Ganson transforms eleven scraps of paper into flying machines resembling a wing-flapping swarm of slow-motion butterflies. In another, a waddling wishbone of a chicken pulls a multiwheeled machine. In yet another, an empty yellow chair is made to meander across a rock. These technological works of art are an absolute joy!

The "Doc Edgarton: Flashes of Inspiration" exhibit tells the story of one of MIT's most beloved professors, Harold "Doc" Edgarton, an electrical engineer who developed the electronic stroboscope. His invention enabled him to take photographs of phenomena that happen much too fast for the human eye to process or the traditional camera to capture. The photographs, many on display here, do nothing less than open up a new window on the world. His best known image is also one of the simplest: the splash of a drop of milk. Images of a swinging golf club, a kicked football, and the aerial gyrations of acrobats, captured the public imagination. The technology found practical applications as well. During World War II, Edgarton developed a method of taking aerial photographs at night, a technique that proved useful in planning the Normandy invasion. During the cold war, he found a way of photographing nuclear explosions. In his later years, Edgarton focused his energy on sonar and underwater photography, working with explorer Jacques Cousteau to bring light to the dark ocean depths.

The final exhibit, "Mind and Hand: The Making of MIT Scientists and Engineers" takes a trip through the educational evolution of MIT from the mid-1800s to the present. Artifacts include old textbooks, instructional lab equipment, problem sets, and slide rules. We learn that in the early 1900s, a merger between MIT and Harvard was proposed but fiercely resisted by MIT faculty, students, and alumni. Displays describe MIT's role in both world wars, with an emphasis on the Radiation Laboratory. You can see some original magnetrons used to produce the radar beams. Look for the display on "hacks," the elaborate, nondestructive practical jokes for which MIT students are famous.

### Visiting Information

The MIT Museum is located at 265 Massachusetts Avenue in Cambridge, Massachusetts. To get there, take the Red Line subway to Central Square and walk down Massachusetts Avenue toward the Charles River. Museum hours are from 10:00 A.M. until 5:00 P.M. daily. Admission is $7.50 for

> Website: www.mit.edu/museum
> Telephone: 617–253–8994

adults, $3.00 for youth ages 5 through 18 and senior citizens. Combine your trip to the museum with a visit to the campus as described in chapter 3.

## National Museum of Nuclear Science and History, Albuquerque, New Mexico

Visit this unique museum to learn all about the history of the Atomic Age, from early nuclear research in the 1920s and 1930s through today's peaceful applications. The museum's eighteen exhibit areas include displays on the pioneers of nuclear science with an emphasis on the contributions of Madame Curie. In another area, you can discover natural and artificial sources of radiation in the world around you. The "Nuclear Medicine" exhibit traces the evolution of nuclear materials in medicine, from a quack medical device called the Revigator to the high-tech Gamma Camera. The

"Power Up" area focuses on nuclear energy. Don't miss the collection of objects that reveal how the atom has made its way into the popular culture.

Most of the museum is dedicated to the history of nuclear weapons, starting with discoveries that eventually led to the Manhattan Project. Displays here tell the story of Oak Ridge, Hanford, and Los Alamos. You can also view a life-size diorama of the Trinity test site. Panels examine the decision to drop the bomb, and accompanying displays feature the atomic attack on Japan. Photos of Hiroshima and Nagasaki can be seen near full-scale mockups of "Fat Man" and "Little Boy." Next come exhibits on the hydrogen bomb and the cold war era, complete with models of nuclear warheads. In the "Submarines, Modern Missiles, and Warheads" area, a Trident missile is dissected into its stages. Outside the building is Heritage Park, a nine-acre area displaying aircraft, missiles, railcars, and nuclear submarines. A B-29 bomber can be seen here, along with a B-52 and a Russian MiG-21. Two antique cars of the type used during the Manhattan Project, a 1941 Packard Clipper and a 1942 Plymouth Special Deluxe, have been restored to their original condition.

### Visiting Information

The museum has recently moved to a new location in southeastern Albuquerque at 601 Eubank Boulevard, SE. The museum is open daily, except for major holidays, from 9:00 A.M. until 5:00 P.M. Admission is $6 for adults, $5 for seniors, and $4 for youth ages 6 through 17.

> **Website:** www.nuclearmuseum.org
> **Telephone:** 505–245–2137

# A STATE-BY-STATE LIST OF SITES

## Illinois (all sites are in or near Chicago)

## Indiana

## Kentucky

## Maine

## Massachusetts

## Michigan

## Minnesota

## Missouri

## Nevada

## New Hampshire

## New Jersey (all sites are in Princeton)

## New Mexico

## New York

## Oregon

## Pennsylvania

## Puerto Rico

## South Dakota

# RESOURCES FOR THE SCIENTIFIC TRAVELER

This book is by no means an exhaustive list of all the physics- and chemistry-related sites across America. If your appetite for scientific travel has been whetted, here are some resources where you can find even more sites to visit.

*The Traveler's Guide to Nuclear Weapons: A Journey Through America's Cold War Battlefields* by James M. Maroncelli and Timothy L. Karpin describes 160 historic sites related to nuclear weapons. The chapters trace the production of nuclear weapons from the mining, refining, and enriching of uranium on through to the testing of the weapons. Sites include factories, laboratories, and proving grounds. The book is available only on CD and can be purchased at www.AtomicTraveler.com.

The American Chemical Society's National Historic Chemical Landmarks Program grants landmark status to places, discoveries, and achievements in chemical history. So far, about sixty landmarks have been designated, and the society adds more each year; however, only a few landmarks are actual sites you can visit. Several sites (Joseph Priestley's house, Berkeley's Gilman Hall, and Columbia's Havemeyer Hall, for example) are listed in this book. Go to the American Chemical Society's website and click on "education" or go directly to the website at http://center.acs.org/landmarks/.

The American Physical Society's Historic Sites Initiative is similar to, though not nearly as well developed as, the program for chemistry. The initiative places a plaque at sites where important events in the history of physics have occurred. Go to the American Physical Society's website and click on "Physics for All" under programs or go directly to the website at http://www.aps.org/programs/outreach/history/historicsites/.

The National Park Services National Historic Landmark Program has hundreds of interesting sites to visit. You can search through the sites by state or by topic. Visit the website at www.nps.gov/history/nhl/.

For scientific travelers visiting Europe, a copy of *The Scientific Traveler: A Guide to the People, Places, & Institutions of Europe* by Charles Tanford and Jacqueline Reynolds should be tucked away in your luggage. Each chapter is dedicated to a different European country and begins with a historical introduction followed by a description of principal places to visit. This wonderful book is out of print, but you may be able to find a copy on the internet.

# INDEX

# ABOUT THE AUTHOR

A native of Paducah, Kentucky, Duane S. Nickell teaches physics at Franklin Central High School in Indianapolis, Indiana, and is an adjunct faculty member at Indiana University/Purdue University at Indianapolis. He holds a bachelors degree from DePauw University, a masters from the University of Kentucky, and a doctorate in education from Indiana University. Dr. Nickell has won numerous teaching awards, including the prestigious Presidential Award for Excellence in Science and Mathematics Teaching, the nation's highest honor for science and mathematics teachers. He is the author of the first book in the Scientific Traveler series, *Guidebook for the Scientific Traveler: Visiting Astronomy and Space Exploration Sites across America.*